HYDROGEN AND ENERGY

ENERGY ALTERNATIVES SERIES
Series editor: C. A. McAuliffe

HYDROGEN AND ENERGY

by

CHARLES A. McAULIFFE

Department of Chemistry,
University of Manchester Institute of Science and Technology,
Manchester, United Kingdom

First published 1980 by
THE MACMILLAN PRESS LTD
London and Basingstoke
Associated companies in Delhi Dublin
Hong Kong Johannesburg Lagos Melbourne
New York Singapore and Tokyo

British Library Cataloguing in Publication Data

McAuliffe, Charles Andrew
   Hydrogen and energy.—(Energy alternatives series).
   1. Hydrogen as fuel
   I. Title   II. Series
   665′.81   TP359.H8

   ISBN 978-1-349-02637-1   ISBN 978-1-349-02635-7 (eBook)
   DOI 10.1007/978-1-349-02635-7

This book is dedicated with love
to Margaret,
and to Amy, Juliette and Andrew

CONTENTS

CONTENTS

PREFACE

This book sets out to do a number of things. First, it attempts to place before the reader a list of energy sources currently available and to describe critically the attractions and disadvantages of each; sources which might be possible, for example fusion power, are also discussed. Secondly, the inevitability of an energy carrier not based on fossil fuels, and not the synthetic energy carrier electricity, leads to the main topic of the book—hydrogen, and its place in the future energy scenario. As an energy carrier hydrogen has many advantages over electricity, the most important being that hydrogen is storable. The possible sources of hydrogen, its transport, storage, uses, and safety aspects are set out at some length, though it is hoped that the general level of discussion will be readily comprehensible to anyone who has completed secondary school science.

I do not accept the proposition that a 'hydrogen economy,' one in which hydrogen is the principle fuel, is inevitable, but I do see that hydrogen will figure very prominently in fuel use in the next century. Other synthetic fuels, the most important of which is methanol, may play an important role, and I have briefly discussed this possibility. I feel very strongly that vastly increased funds for research and development of coal gasification and liquefaction need to be provided in the Western world, and I have devoted a considerable part of the book to making the point that coal can provide the necessary bridge between the hydrocarbon era and, for instance, the solar energy era. In making coal more acceptable as a fuel large amounts of hydrogen are necessary, and thus I see the main impetus for an eventual 'semi-hydrogen economy' coming from the need for vastly increased hydrogen production to convert coal to hydrocarbons.

Since a fusion reactor, the sun, is readily available to us I would prefer to see the world continue to rely on solar energy in the future as it has done in the past. Apart from this bias for solar energy and against nuclear power sources I think the book is essentially free from other biases and I have attempted to present the facts as I have been able to obtain them.

Readers of literature devoted to the subject of energy will be aware of the abundance of different units. Little can be done about this except to refer readers of this book to the appendix on energy units.

I would like to thank my wife, Margaret, for the patience and interest she has shown in my own interest in the subject of hydrogen as a fuel. I am also most grateful to my typist, Joan Colclough, for an excellent job of manuscript production.

Manchester, 1980   C. A. McA.

CHAPTER 1

ENERGY—BACKGROUND CONSIDERATIONS

The year 1973 is an important one in the history of Man's use of energy. The Arab oil embargo initiated a period of deep concern by the governments of many industrialised nations; rapid price rises, in addition to uncertainty of security of supply, have forced oil-importing nations to reassess energy policy. This initial reassessment involved two main factors: first, to make energy use more efficient and secondly to develop national energy resources to maximum capacity. The latter intention has certainly gripped the imagination and determination of Western nations, but as for more efficient use of energy, little has been heard of this since the oil embargo ended. Certainly the UK Department of Energy's feeble advertising campaign exhorting more efficient use of energy is no excuse for an almost complete lack of legislation enforcing this vital matter. One can recall Emmanuel Shinwell, shortly after World War II, urging that no houses should be built without adequate insulation against heat losses. Thirty years later there is still no such legislation. This book is intended to offer information about the increased incorporation of hydrogen into the overall energy economy and, as such, does not deal explicitly with political matters. This author does not subscribe to the interpretation of the phrase 'energy crisis' as meaning energy shortage. There is no shortage of energy, but we must adapt quickly to changing patterns of energy supply. One hopes that the politicians are up to the challenge—had it not been for the oil embargo of 1973, I fear that western governments would still be blissfully ignoring the fact that gaseous and liquid hydrocarbon reserves are rapidly running out.

It is a direct consequence of plentiful and cheap energy that present day society makes such wasteful use of it. The design of commercial and residential buildings which ignores energy conservation has already been mentioned, but the design and size of private cars, transport of goods by road instead of by rail and water, and the increasing use of materials synthesised from petroleum are further examples. Typical of our misunderstanding of energy economics is the approval society presently gives to the trend towards cars with smaller engines and yet society completely ignores the fact that cars are not built to last—what about the enormous quantities of energy which go into fabricating cars of diminished life expectancy? It is not likely that much can be done in the short-term to alter such trends, but programmes of education and legislation can help improve the intimate connection between the aspirations and life-styles of people and energy use. At present, opportunities to improve the efficiency of energy consuming and converting equipment are retarded by short-term economic considerations which ignore the longer term impact on energy resources.

Introduction

The decision of OPEC to raise the price of oil and curb supply will, in the future, be seen as the most important factor in prolonging fossil-fuel supply. Had the energy situation of the last decade been allowed to continue unchecked, then oil reserves would have dried up perhaps at the turn of the century. The price increase mechanism has initiated what will hopefully become an intensive research and development programme in the in-dustrialised countries in the field of energy. This includes the stepping up of the worldwide search for new oil deposits in remote areas, improving the efficiency of recovery methods (presently less than half the oil in a well is recovered), and, most important, searching for new sources of energy.

Oil has achieved its position as the main source of world energy through a combination of properties such as cheapness and convenience and flexibility of use. It has been generally considered as irreplaceable as a fuel for the internal combustion engine. The degree of dependence on oil in the world energy balance is still growing, and it is increasingly difficult to see how such growth can be modified significantly or reversed in the next 25 years. This is due to the long lead times required on the supply side of the energy equation to bring into full production those other fuels which could take the pressures off the conventional fuels; or on the demand side, to adapt fuel technology or otherwise reduce the volume required. It has been projected that by 1980 the USA will be dependent on foreign sources for half its supply of oil, assuming current trends apply. To release auxiliary quantities of energy on the necessary scale from nuclear power, coal, or non-conventional oil supplies will involve research and development, scaling up and plant construction periods plus the time required to modify associated equipment and practices in energy consumption which stretches into a ten year span. With this fact in mind former-President Nixon's goal of energy independence for the USA by 1980 has proved unrealistic.

The fact that mankind has used in the last thirty years more energy than was used in its previously recorded history is somewhat frightening. It is estimated that the same amount of energy will be used in the next 15 years, and the same again in the following seven years. Energy demand is currently increasing at a rate of about 5 % per annum (figure 1.1). The imbalance of worldwide consumption is also significant: for example, the USA with 6 % of the world's population uses 34 % of consumed energy (see table 1.1), whilst an area such as the Far East containing 28 % of the world's population accounts for 3 % of the world's energy consumption.

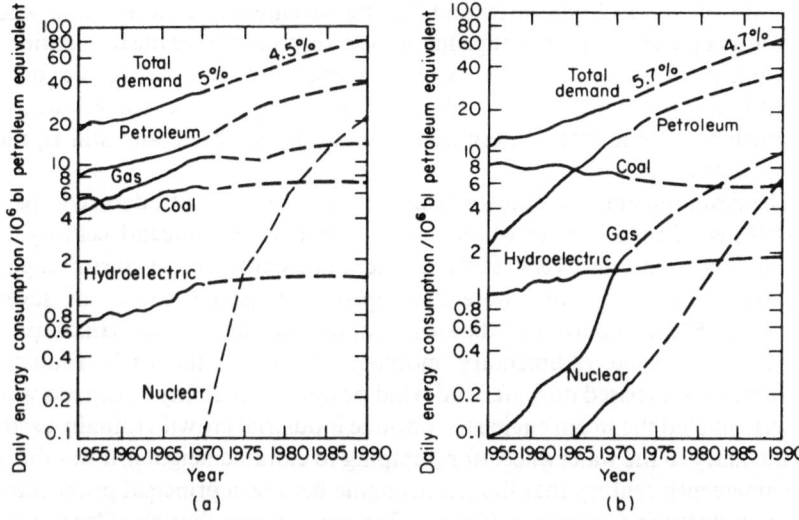

Figure 1.1 An evaluation of recent and projected annual consumptions of primary energy resources in (a) the USA and (b) Western Europe. The % figures above the top curves indicate rate of growth of consumption. (Reproduced from reference 1 with permission)

Table 1.1 Estimates of energy consumption. $1 \, Q \equiv 10^{18} \, \text{BTU} \equiv 1.05 \times 10^{21} \, \text{J}$
$\equiv 2.93 \times 10^{14} \, \text{kWh(th)}$ (y, year; p, person)

USA	(1970)	$0.07 \, Q/y \, [2 \times 10^8$	people, 11.7 kW(th)/p]
	(2000)	$0.16 \, Q/y \, [3 \times 10^8$	people, 17.8 kW(th)/p]
	(2020)	$0.3 \, Q/y \, [4 \times 10^8$	people, 25 kW(th)/p]
World	(1970)	$0.24 \, Q/y \, [4 \times 10^9$	people, 2 kW(th)/p]
	(2000)	$2.1 \, Q/y \, [7 \times 10^9$	people, 10 kW(th)/p]
	(2050)	$6 \quad Q/y \, [10 \times 10^9$	people, 20 kW(th)/p]

The manipulation of energy has been an essential component of man's ability to survive and to develop socially. Primitive peoples and animals adapt to a changing environment, but it is a unique characteristic of modern man that he alters his environment to suit his own needs. Use of energy has been the key to supply of food, physical comfort, and quality of life beyond crude survival. Availability of resources and the technology to convert them into an acceptable form are the two factors which limit the extent to which energy can be utilised. Resources have long been generally available but devices for conversion have been a fairly recent development. It is interesting to trace such a development and to compare it with the improvement in the quality of life.

Water power for irrigation purposes, exploiting gravitational potential energy, was known from early times. The horizontal waterwheel appearing in about the first century BC had a power capacity of perhaps 0.3 kilowatts. The fourth century saw the introduction of the vertical waterwheel having a capacity of about 2 kilowatts, and by the seventeenth century waterwheel power was the most important prime-mover, responsible for the foundation of the industrial revolution in Western Europe. Windmills appeared in the twelfth century having a respectable capacity ranging from several kilowatts to as much as 12 kilowatts. Intermittent operation has been, and still is, their greatest drawback.

The development of the steam prime mover is relatively modern, compared with the windmill and waterwheel; it was not until the seventeenth century that steam was used effectively. Early in the eighteenth century steam engines involving moving pistons were developed as a power source of several kilowatts. More importantly, the steam engine was the first mechanical prime mover to provide rudimentary mobility. Although the early Industrial Revolution was based on water and wind power, the inability to expand water power enabled the steam engine to continue industrial growth. Initially used as an auxiliary to the waterwheel (for pumping to elevated heights), it was during the nineteenth century that the steam engine became a principal prime mover for manufacturing industries. Figure 1.2 shows the contribution of large power machines to the social development of man since 1700. Of significance is that power output of energy-conversion devices has increased roughly 10 000

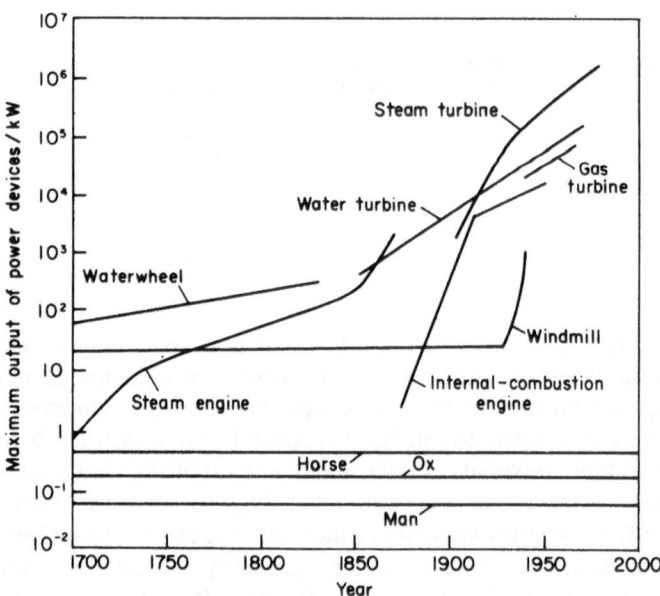

Figure 1.2 Power output of basic machines has risen by more than five orders of magnitude since the start of the industrial revolution (ca. 1750). For the steam engine and its successor, the steam turbine, the increase in output has been more than six orders, from less than a kilowatt to more than a million. All are surpassed by the largest liquid fuel rockets (not shown) which for brief periods can deliver more than 16 million kilowatts. (Reproduced from reference 2 with permission)

times, and since most of the growth has taken place within the past century, we are only now experiencing the combined results.

In order to examine options for the future there is much to be learnt from past trends. Energy consumption in the USA has been well documented and useful lessons concerning 'energy-use revolutions' are available. During the past century consumption in the USA has increased eighteen fold. Fuel wood provided almost all energy in 1850 (figure 1.3), but by 1910 coal accounted for

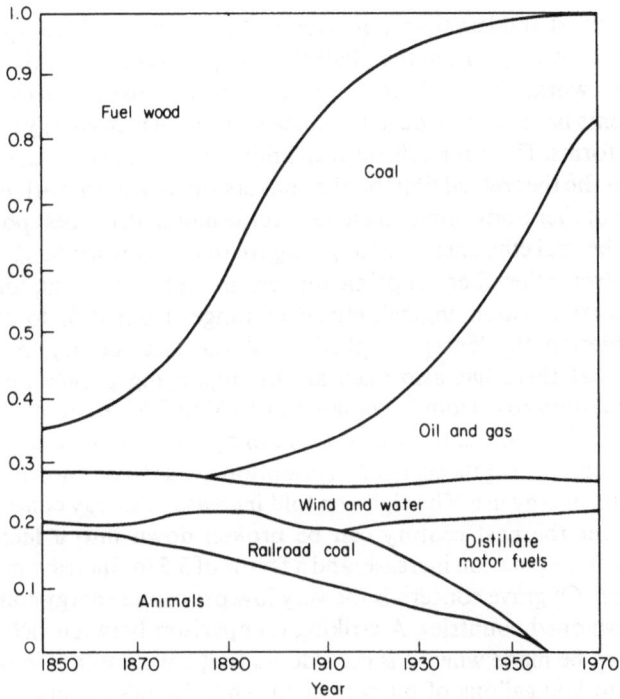

Figure 1.3 Segmentation of the fuel input of the USA, 1950–1970

about 75 % of the total energy consumption and fuel wood had declined to some 10 %. In the 50 years between 1910 and 1960 the dominance of coal was supplanted by that of oil and natural gas and in 1970 the latter sources accounted for 76 % of the energy consumed in the USA, compared to 61 % worldwide. Table 1.2 contains an interesting synopsis of energy input into the UK over recent years.

Approximately 50 years seem to have been needed for the energy economy to shift substantially to a new fuel. This is determined primarily by the operating life-time of power machinery and secondly by the long lead time for producing available manufacturing and supply capability.

A century ago energy resources were mainly used to produce heat for

Hydrogen and Energy

Table 1.2 United Kingdom energy input [UK Digest of
Energy Statistics (HMSO 1975, 1979)]

Energy type	% 1963	% 1973	% 1978
Coal	68.4	37.9	35.3
Petroleum	30.0	46.0	41.0
Nuclear/hydroelectric	1.5	3.5	4.5
Natural gas	0.1	12.6	19.2

physical comfort and less than a quarter of the heat was utilised for industrial
activity, but nowadays more than half the energy consumed in the West goes
into useful work. As well as a change in end use of energy, steady
improvements have been made in the efficiency with which energy is converted
into useful forms. There is no theoretical limit to the efficiency of energy use for
heating, but the theoretical limit on the conversion of heat to work is governed
by the Carnot thermodynamic efficiency. At the moment, the best power plants
operate at thermal efficiencies of 40%, a figure that may reach 50% by the year
2000. However, other thermal prime movers are not so efficient; for example,
the internal combustion engine's efficiency ranges from 10% to 25%.

In considering the 'energy explosion' of the last century, it is well to
remember that there has also been an accompanying population increase.
World population rose from 1.6 billion* in 1900 to 3.5 billion in 1970 and on
current trends is expected to reach 8.7 billion by the year 2020. But, as well as
population size, per capita energy consumption needs to be known in order to
evaluate total energy use. The eighteen fold increase in energy consumption in
the USA over the past century can be broken down into a factor of five
attributable to population increase, and a factor of 3.5 for increase in per capita
consumption. Of grave concern is the very low per capita energy consumption
of underdeveloped countries. A striking comparison between rich and poor
countries can be made when it is considered that consumption in the USA is
equivalent to 900 gallons of oil per capita while India's is only 9.

An approximate linear relationship exists between per capita energy
consumption and gross national product (figure 1.4). Underdeveloped
countries, which contain the majority of the world's population, are still
primarily engaged in maintaining a level of subsistence. Power resources for
the transition to a literate, industrial society are not yet available.

Since 1900, the average per capita energy consumption in the industrialised
West has increased at a rate equivalent to a doubling every 50 years. There is
every likelihood of this trend continuing for several decades. In spite of
growing public concern with the impacts of such growth, there is actually very
little that can be done to limit it, other than through scarcity or rationing.

It is conceivable that in the next century per capita energy consumption in
the advanced countries will approach an equilibrium level. The quality of life

---

* In this book one billion = 10^9.

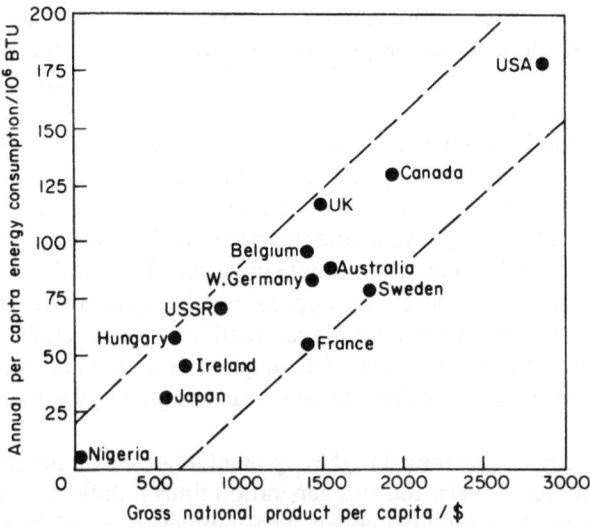

Figure 1.4 Correlation between commercial energy use and gross national product. (Data taken from reference 3 with permission)

for the majority of the population will be less dependent on increased energy use and, moreover, environmental constraints may make energy use more costly, encouraging increased efficiency in end use. Also population equilibrium in advanced countries may lead to an eventual levelling off of total energy need.

America's share of the world energy consumption is likely to drop from its present 34% to about 25% by the year 2000, because of relative population increases in the rest of the world. Per capita increase in consumption in the USA is currently running at 1% per annum, and worldwide per capita energy consumption is increasing at the rate of approximately 1.3% per annum. It may well be another century before world average per capita energy consumption even approaches current levels in the USA. At that time, the energy gap between the USA and underdeveloped countries will still be large. With unaltered trends it would take 300 years to close the gap. By the year 2000 the world's average per capita energy consumption will have moved only from the present one fifth of the USA average to about one third. If developing parts of the world reached present US standards of living by 2000, worldwide energy consumption would increase roughly ten fold. Whilst such improvements must be viewed as socially desirable one must conclude that the industrial base does not exist for such a change; the strains on resources other than energy would be too great.

The energy resources of the earth

Of primary sources, solar energy supplies by far the major part of the energy of the world. This source is supplemented by small amounts of heat from the earth's interior and by tidal energy from the gravitational system of the earth, moon, and sun. In addition to these is that energy obtained from the natural radioactive elements distributed on the earth's surface.

Currently, mankind is dependent almost exclusively for his energy supply on solar energy, stored millions of years ago in the form of fossil fuels. Plant leaves capture a small fraction of incident solar radiation and store it chemically by the mechanism of photosynthesis, and this becomes the energy supply essential for the existence of the plant and animal kingdoms. Biologically stored energy is released by oxidation at a rate approximately equal to the rate of new storage. Over millions of years, a residual fraction of the vegetable and animal matter is burned under conditions of incomplete oxidation and decay, thereby giving rise to the fossil fuels that provide most of the energy for industrialised societies.

We have become accustomed to the exponential growth in the consumption of energy from fossil fuels, and this generation finds it difficult to realise how quickly will pass the fossil fuels epoch. The complete cycle of the exploitation of the world's fossil fuels will involve only about 1300 years with the principal segment* covering only about 300 years. It is hoped that nuclear energy or solar energy capture will be developed to levels of sophistication so that economic expansion can be sustained at the present rate.

It is useful to put an order of magnitude on these resources. Measurements have shown that the total solar radiation intercepted by the earth's diametric plane is of the order of 1.73×10^{17} watts, or an average of 1.395 kilowatts per square metre (figure 1.5). The average rate of heat flow from the interior of the earth's surface has been put at 0.063 watts per square metre. This figure is based on the measurements of geothermal gradients as well as on the conductivity of the rocks (or seawater) involved. For the entire earth's surface, this means an energy output of some 32×10^{12} watts. Rates of heat convection by hot springs and volcanoes amount to only about 1% of the rate of conduction. Energy from tidal sources has been estimated at 3×10^{12} watts. From the above figures it can be seen that solar radiation accounts by far for the largest energy input to the earth (94–98% of it). In fact, the amount of solar energy irradiated onto the earth's surface in about a day is equivalent to all the energy used by mankind to date!

Of more importance is the proportion of this huge energy influx which can be exploited. About 30% of the radiation is directly reflected and scattered back into space as short-wavelength radiation, whilst another 47% is absorbed by the atmosphere, the land surface and the oceans, and is converted directly into heat at the ambient surface temperature. Evaporation, convection, precipitation, and surface run-off of water in the hydrologic cycle takes up the

---

* Principal segment is defined as the period during which all but the first 10% and the last 10% of the fuels are extracted and burned.

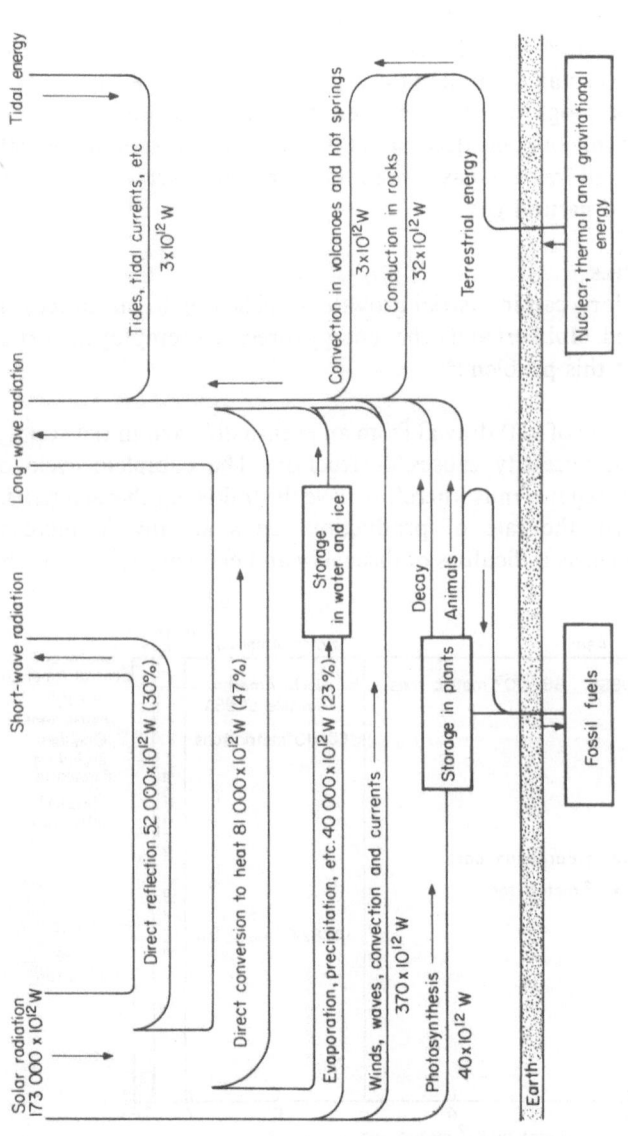

Figure 1.5 Flow of energy to and from the earth is depicted by means of bands and lines that, by their width, suggest the contribution of each item to the earth's energy budget. The principal inputs are solar radiation, tidal energy, and the energy from nuclear, thermal, and gravitational sources. More than 99% of the input is solar radiation. The apportionment of incoming solar radiation is indicated by the horizontal bands beginning with 'direct reflection' and reading downwards. The smallest portion goes to photosynthesis. (Reproduced from reference 2 with permission)

remaining 23%. A small fraction (0.2%) drives the atmospheric and oceanic convection and circulation which is eventually dissipated as heat by friction. Finally, an even smaller fraction (0.02%) is captured by the chlorophyll of plant leaves, where it becomes the essential energy supply of the photosynthesis process and eventually of the plant and animal kingdom.

Photosynthesis fixes carbon in the leaf and stores solar energy in the form of carbohydrates. A very small fraction of the organic matter produced, however, is deposited in peat bogs or other oxygen-deficient environments under conditions that prevent complete decay and loss of energy. During the last 600 million years this small fraction has turned out to be our reserves of coal, oil shale, petroleum and natural gas.

Demand and reserves

The question: how long can industrial growth be fuelled by fossil sources? is now frequently asked. Hubbert and other energy forecasters employ historical records to highlight this problem:[3]

Exploitation consists of withdrawal from an eventually fixed initial supply, and utilisation of necessity causes destruction. The complete cycle of exploitation of a fossil fuel must therefore have the following characteristics. Beginning at zero, the rate of production tends initially to increase exponentially. Then, as difficulties of discovery and extraction increase, the

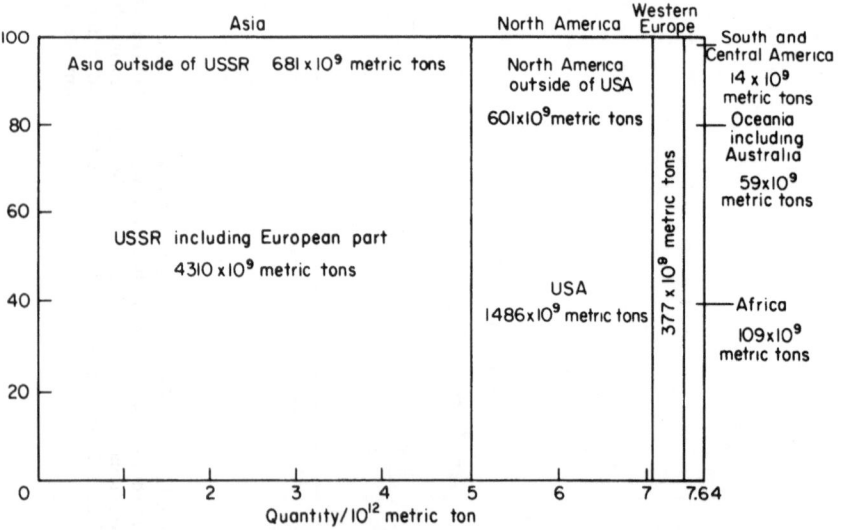

Figure 1.6 Coal resources of the world derived from data of Averitt, US Geological Survey. The figures represent the total initial resources of minable coal—defined as 50% of coal present. The vertical columns show the apportionment of coal among the continents. (Reproduced from reference 2 with permission)

production rate slows its growth, passes one maximum or more and, as the resource is progressively depleted, declines eventually to zero.

If known past and prospective future rates of production are combined with a reasonable estimate of the amount of a fuel present, one can calculate the possible length of time that the fuel can be exploited. In the case of coal, reasonably good estimates of the amount present in given regions can be made on the basis of geological mapping and a few widely spaced drill holes, in as much as coal is found in stratified beds or seams that are known over extensive areas. Such studies have been made on all the coal-bearing areas of the world.

Figure 1.6 shows a recent compilation of world coal reserves made by Paul Averitt of the US Geological Survey. His figures represent minable coal, here defined as 50% of the coal actually present. Included is coal in beds as thin as 34 centimetres and extending to depths of 150 metres or, in a few cases, 2000 metres. Averitt estimates an initial supply of $7-6 \times 10^{12}$ metric tons and assuming that the present production rate of three billion metric tons per year does not double more than three times, one can expect that the peak in the rate of production will be reached sometime between 2100 and 2150. Disregarding the long time required to produce the first 10% and the last 10%, the length of time required to produce the middle 80% will be roughly the 300 year period from 2000 to 2300 (figure 1.7).

It is much more difficult to make estimates for gaseous and liquid hydrocarbons than it is for coal because, as Hubbert points out, these deposits occur in fairly limited areas in sedimentary basins and at depths which can vary between 8 m and 8 km. However, once an area is under exploitation it becomes possible to make estimates, and it is even possible to extrapolate to areas where oil has not yet even been discovered.

It is becoming uncomfortably clear that new finds of gas and oil will be in areas in which exploitation is difficult, Alaska, the North Sea, and even in the

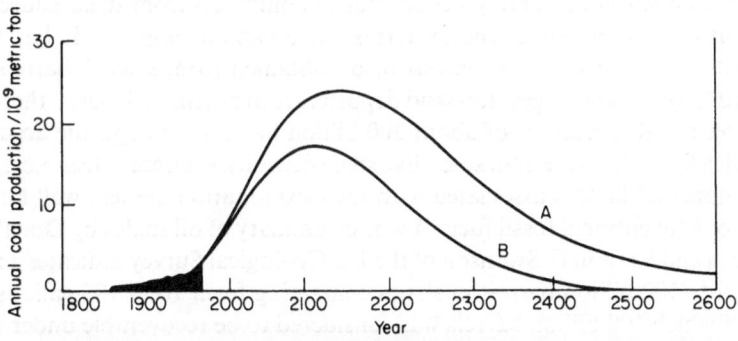

Figure 1.7 The cycle of world coal production derived from estimates of supply and production by (A) Averitt and (B) Hubbert. Total supply A amounts to 7.6 $\times 10^{12}$ metric tons, and B to 4.3×10^{12} metric tons

relatively shallow waters of the Gulf of Mexico. Even on continental USA the amount of oil discovered per foot of exploratory drilling has fallen from 84 bl/ft in the 1930s to about 11 bl/ft during the last decade (bl = barrel). Figure 1.8 contains estimates, by two respected energy reserve pundits, of the cycle of world oil production; the middle 80 % of production lasting either an estimated 58 (Ryman) or 64 (Hubbert) years.

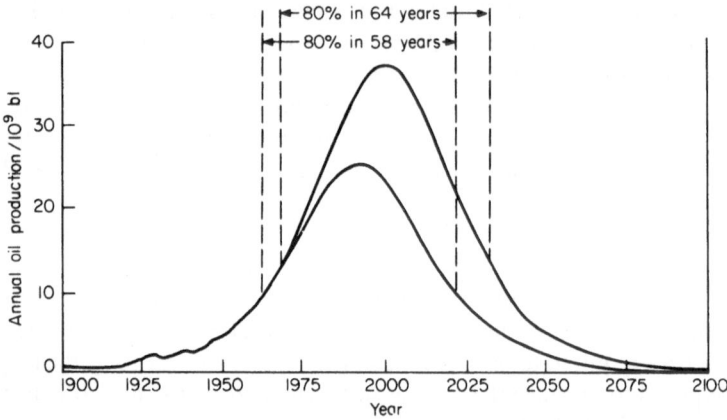

Figure 1.8 The cycle of world oil production derived from estimates of supply and production by (A) Ryman and (B) Hubbert. Total supply A amounts to 2100 $\times 10^9$ bl and B to 1350×10^9 bl

The prognosis for natural gas is no more comforting either. Whilst known world reserves amount to between 34 and 48×10^{12} cubic feet, potential future recoverable reserves have been put at about $90–340 \times 10^{12}$ cubic feet. This figure gives a projected life-time for natural gas, at predicted future consumption rates, of about 40 years.

A substantial amount of oil can be extracted from tar sands and oil shales, where production has barely begun. But obtaining oil from these sources is difficult and expensive in energy terms—one cannot pump it. It has been estimated that for every 10 barrels of oil obtained from shale 7 barrels are needed to get it. The largest tar-sand deposits are in northern Alberta; there are total recoverable reserves of about 300 billion barrels although the accuracy of such a figure is less certain since, like shale deposits, economic, technical, and environmental factors associated with their exploitation are less well known than for conventional fossil fuels. A world summary of oil shales by Donald C. Duncan and Vernon F. Swanson of the US Geological Survey indicates a total of about 3100 billion barrels in shales containing from 10 to 100 gallons per ton, of which 190 billion barrels was considered to be recoverable under 1965 conditions.

Non-fossil energy use

Assuming the premise, that at sometime in the future fossil fuel reserves will be completely exhausted, the question naturally arises of how alternative energy sources will be exploited? As well as solar energy these sources include tidal energy, geothermal energy, and nuclear energy. Although the former is, at present, by far the largest source, the potential for nuclear energy exploitation is far greater than that of the rest. Quantifying the amount of nuclear energy available for mankind is synonymous with forecasting the improvement in technology for its conversion into more usable forms.

Certainly there is enough solar energy incident on the earth's surface to sustain economic growth for thousands of years, but the problem lies in its conversion to forms which can be beneficial to man.

The use of solar energy for electricity generation and heat generation has been examined for a number of years. The thermodynamic potential of solar energy is the same as that for a black body at 6000 K. The solar energy falling onto a surface that is perpendicular to the sun's rays is, however, only 1.3 kW/m² before atmospheric absorption (which can amount to 30%). This is a very low flux density, about one five-hundredth of that on the heating surfaces of a modern steam boiler.

Nowadays there is little direct use of solar power; what there is amounts to small scale use such as heating works and generating electricity for spacecraft by means of photovoltaic cells. Much more substantial installations will be needed if solar power is to replace fossil fuels on an industrial scale. Such a scheme would not only require very large areas of land for collecting of required amounts of energy, but also suffer from the disadvantage of intermittent supplies of radiation. Hence because supply and demand for energy would never perfectly match, some provision for large scale storage would be required (but see later how a solar–hydrogen economy overcomes this problem).

The most attractive scheme for solar energy utilisation is the concept of the solar house. In fact, solar water heaters are already in commercial use in many places. For space heating, the solar collector is typically a black metal surface that readily absorbs sunlight; it is covered with one to three panes of glass to reduce heat loss. The glass transmits incoming sunlight, but absorbs the longer wavelength radiation emitted by the hot metal, so that a greenhouse effect is created and the effectiveness of the collector is increased. The heat is held by water or air that circulates through the collector during the day, and part of it is stored for release at night. Hot water, hot rock, and chemical (change of phase) storage systems have been experimentally tested. The major drawback is the efficiency with which the solar energy can be converted into heat.

Solar cells, which convert sunlight directly into electricity, are the predominant source of power for space satellites. However, high manufacturing costs and short lifetime means that they are not competitive with other means for generating electricity for terrestrial use. Moreover the best cells produced today, using silicon semiconductors, show conversion efficiency of only 13 % whilst developing a potential of a mere 0.5 volts.

The most favourable sites for developing potential solar power are areas not more than 35 % degrees north or south of the equator. Again this creates a major problem of how to transport the energy to the main consuming areas of the world, and this brings to mind the use of hydrogen as an energy carrier.

Indirect use of solar energy through water power has mainly been exploited only where geographic conditions are favourable for the construction of reservoirs. This means that energy can be stored fairly efficiently and subsequently only used for peak-shaving periods. Suitable sites for exploitation in developed countries are now limited so that hydro-electric power production in them will not increase significantly its present 3–4 % of energy supply. However, the case for underdeveloped countries is more promising. Tidal power again suffers from siting problems, and will probably only make a minor contribution as a new energy source. The use of tidal energy is ecologically attractive, but as table 1.3 shows there are relatively trivial amounts of energy in non-solar sources, and money for research should be invested in solar energy capture rather than such unimportant sources as the tides.

Geothermal heat has been described as a form of fossil nuclear energy, since it is produced primarily by the decay of radioactive materials within the earth's interior. Radiative and conductive processes transport small amounts of heat to the surface, but large deposits of heat within the earth's crust are apparently the result of geologically recent intrusions of matter rock from the mantle. Only volcanic sources are significantly exploited at present. In all cases the heat is used to generate steam for producing electricity. Of the three main areas worthwhile for geothermal-power production, the one in northern California has the highest capacity. Whether dry geothermal resources can be economically competitive with other sources of energy will depend in part on how deep the deposits are, since the expense of drilling through hard rock is expected to be a major cost. The amount of energy available is probably comparable with that in the tides, but the technology needed to extract geothermal heat is much less involved.

Table 1.3 Available non-hydrocarbon energy sources (from M. K. Hubbert, *Scientific American*, Sept. 1971)

Source	Potential/10^9 W	
Solar	177 000 000	(equivalent to 10^5 × present installed electric power capacity
Water	2 900	
Tidal	64	
Geothermal	10	
Nuclear	very large if breeder reactors used or if fusion becomes commercial	

Nuclear energy has emerged as a major addition to traditional energy sources and the only new technology developed to the point of commercial use in the past 30 years. However, there are very grave hazards involved in largescale production of energy by nuclear means. Moreover, the immense capital costs involved in a large-scale nuclear programme need to be weighed against the less costly, environmentally more attractive options such as large-scale coal gasification programmes.

Nuclear power must be considered under the two headings of fission and fusion. Fission involves the splitting of nuclei of heavy elements such as uranium. Fusion involves the combination of light nuclei such as deuterium. In either case the reaction causes the conversion of matter into energy. Uranium-235, which is a rare isotope, is the only atomic species capable of fissioning under relatively mild environmental conditions. Since natural uranium consists of some 99 % non-fissionable uranium-238, the nuclear-fuel epoch would be brief if nuclear-energy was entirely dependent on uranium-235. By breeding, however, wherein by absorbing neutrons in a nuclear reaction uranium-238 is transformed into fissionable plutonium-239 or thorium-232 becomes fissionable uranium-233, it is possible to create more nuclear fuel than is consumed. With breeding, the entire supply of natural uranium and thorium would become available as a fuel for fission reactors.

The very recent origins of the nuclear power industry are reflected in the fact that at the end of 1968 only 54 power reactors were in service in the non-communist countries. Their combined capacity was 9800 MW, against a total generating capacity of nearly 700 000 MW; five years earlier the comparable figure was only 4493 MW of nuclear capacity. The total stock of nuclear plants in operation in 1968 was thus very small in relation to both the total stock of electricity generating capacity and annual additions to it. The upsurge in nuclear ordering since 1966 has been such that about 280 reactors are now in service, with a total capacity of approximately 167 300 MW.

The development of nuclear power in the 1980s is surrounded by considerable uncertainty. Three recent projections of the growth of nuclear power in the 1980s have been made. These ranges do not convey the full extent of the uncertainties: the projections could be set back for example by one serious nuclear failure or by increasing public concern about the safety and siting of nuclear plant. Subject to this qualification, the estimates suggest that 37 countries will have at least one power reactor in operation by 1990: but the same nine countries as in 1977 will still account for almost nine-tenths of the world nuclear generating capacity in 1990, and the four major countries (USA, Japan, UK and West Germany) will still account for three quarters of the total.

In order to reflect economies of scale the size of nuclear reactors will certainly increase. For example, the average size of the reactor installed in 1969 was 361 MW whilst by 1977 this figure will have more than doubled to 771 MW. This is particularly true in the USA where reactors in the

950–1200 MW class account for half the nuclear generating capacity entering service over the next few years.

Nuclear fuels have an even more complex supply structure than nuclear plants. The supply of nuclear fuel is best regarded as a cycle in which the nuclear reactor partially burns uranium which has previously been through various conversion and fabrication stages. After irradiation inside the reactor, the spent fuel elements are reprocessed to separate out the remaining uranium, which is used again. *The radioactive waste materials require long-term storage.*

The complex fuel cycle gives rise to many problems—particularly because of the close connection between the civil and military use of the nuclear materials involved. Owing to the military implications, governments have, so far, retained control over enrichment services and know-how. In the non-communist countries there are only five such plants—one in the UK (Capenhurst), one in France (Pierrelatte) and three in the USA (Oak Ridge, Paducah, and Portsmouth).

D. J. Rose[4] has focussed on AD 2000 to make demand projections on, as plants planned today will then be halfway through their operating lifetime. AEC in the USA estimates 1 200 000 MW of nuclear capacity installed by that date. Beyond the turn of the century, the rate of growth is more difficult to predict: acceptable sites will become very hard to find. The nuclear capacity projected is imagined to be perhaps two-thirds of the total electricity capacity predicted for AD 2000, about four times the capacity of present installations. Various scenarios predict a total cumulative requirement of about 2.5 million tons of uranium oxide (U_3O_8) by AD 2000 and 4 to 4.5 million tons by AD 2010. The precise amount depends on the mix of reactor types, on when the breeder reactors are introduced, and on the actual future electric demand.

The concentration of uranium oxide in mined ore varies (high-grade ore containing upwards of 1600 parts per million oxide) and its price reflects the cost of extraction. High-grade resources (up to $10 per lb) would generate enough power for 30 years if used in thermal reactors and based on present-day demand rates. Growing nuclear demand will, however, mean that higher priced uranium will have to be used. Use of fuel costing more than $100 per lb would eradicate the present nuclear cost advantage. This, then, is essentially the argument which promotes the rapid development of the breeder reactor. In effect, not only are the nuclear energy resources multiplied by a factor of 100 but the resource cost per energy unit drops similarly. To a good approximation, the cost of electricity from nuclear breeders becomes independent of uranium prices.

AEC budgeted $357 million for the development of the liquid metal fast breeder reaction (LMFBR) in 1975 and plans to invest a total of $2556 million from 1975 to 1979. The first demonstration plant is to be built in Tennessee producing 300 MW of electricity. Meanwhile the Soviet Union and France have prototype demonstration LMFBRs already operating and the UK will soon follow.

The LMFBR has substantial technical points in its favour, besides the eventual advantage of resource conservation. It is non-pressurised, and in that respect more completely sealed, simple, and safe. Operating at higher temperatures than conventional light water reactors, it has a projected efficiency of 41 %, comparable to the best fossil-fuel plant. The high thermal conductivity and heat capacity of its coolant—liquid sodium—make it virtually immune to damage in case of mechanical failure of the cooling system external to the reactor. However, the use of liquid sodium creates difficult engineering problems which must be effectively solved before production on a large-scale is envisaged.

If the fusion of hydrogen, such as takes place in the sun, could be harnessed on earth, then seawater could provide enough fuel for the world's energy needs for millions of years. (But one might reasonably ask the question: Since the world has a proven fusion reactor available already—the sun—and it is located at a safe distance, 92.8 million miles, from centres of population, why then not spend money on ways of capturing, storing and transporting this energy, rather than invest money on developing 'suns' on this planet? Such questions seem to be beyond the thought processes of the politicians who are making these immensely important decisions for future generations.)

Fusion was first developed in the hydrogen bomb, where energy is released in an uncontrolled fashion. At one time it was thought that research on a fusion reactor might proceed so quickly that fusion would become an alternative to the first generation of breeder fission reactors. But early projections were too optimistic. No one knew in the early 1950s how slow progress towards a fusion reactor would be because few scientists realised that first it would be necessary to unravel and master the details of a whole new field of science—plasma physics. Commercial fusion power is at least 50 years away.

The problem is to heat up a mixture of deuterium and tritium to a temperature between 10^8 and 10^9 K at a density high enough and for a time long enough that the product of the two quantities exceeds approximately 3×10^{14} atom cm^{-3} s^{-1}. One method relies on confining the highly ionised plasma with strong magnetic fields, say 50 000 to 100 000 gauss, at a density of 10^{14} atom cm^{-3} for a few seconds. The most successful example of this technique so far is the Tokamak, wherein the plasma is confined and heated in a toroidal magnetic field. Deuterium fuel is plentiful in seawater but tritium would have to be bred from lithium in the fusion reactor. Whilst requiring even higher temperatures the deuterium–deuterium reaction will be an attractive goal because of the increased amount of fuel supply.

Soviet researchers demonstrated in 1968 that a fusion reaction could be laser-initiated. The big lasers necessary to test the feasibility of laser fusion are just now becoming available, and lasers for a practical power plant would require much more development. Laser fusion depends critically on the laser pulse ablating a deuterium–tritium target pellet so fast and so evenly that the reaction forces on its surface compress it by a volumetric factor exceeding

1000. Then it undergoes nuclear fusion in about 10^{-12} seconds.

Controlled fusion has, during this 'scientific age', been the most challenging and difficult of all such assignments ever given to physical scientists. The goal of obtaining a net output of energy is still far from being a reality. This fact, along with a host of other technical problems, strongly suggests that AEC's implied goal of beneficial installation after 1995 seems optimistic.

However, having mentioned the severe technological drawbacks associated with commercial fusion it is well worth while to mention its outstanding attractions as a future energy source—attractions which will cause its pursuit to be undertaken with increased endeavour in years to come. On the fuel supply side, deuterium is sufficient for 10^{10} years and lithium probably adequate for any technological age to come. As for safety, the only appreciable radiation hazard is from tritium, which is less hazardous than plutonium by many orders of magnitude. Also the reactor's structure, while surely made radioactive by 14-MeV neutrons, is not liable to pose any appreciable hazard.

What is not generally recognised in the current energy situation is that the eventual solution will depend upon not only developing alternative sources of energy but also on devising new methods for energy conversion. There is more than enough energy available, but either it is not in a form convenient for immediate use or it is located too far away from consuming centres. If historical trends in energy consumption continue, then two primary sources are likely to assume over-riding importance: namely coal and nuclear energy.

If a change from a fossil-fuel is to be made it must be a smooth one, and many potential problems are obvious. For any energy source to be well exploited, it must possess two inherent properties: first, it must be easily transportable to consumer areas and secondly it must be storable to allow for variable demand. Furthermore transport and storage costs should not be a high proportion of overall costs. No problems are encountered with the present-day fuel mix of mainly fossil-fuels, but the longer term picture gives cause for concern. Conversion of nuclear energy, as we know it today, is subject to many contrasts. The heat energy it produces must be converted to a form suitable for transportation. Nuclear power can only be converted economically in large units of 1000 MW or more and furthermore stations, because of their large cooling requirements and because of environmental considerations, must be sited away from population centres.

On the other hand, energy demand is periodic. Energy is required in high-density areas, such as large cities, where natural cooling capacity is already strained and which cannot provide the best location for nuclear power stations. Fear of radio-active fallout also means that nuclear stations must be sited well away from high consumption areas. Furthermore, transport is a heavy user of energy, which means that a readily portable source is essential.

The traditional method of energy conversion is to generate electricity which, on account of a number of properties, is attractive for end use. It is *clean* and easily adaptable for many consumer utilities. Although, only accounting for some 10% of energy end-use in the USA (1970 figures), the demand for electricity is growing at a rate faster than the total energy growth by a factor of two. It is the same in most countries and the growth rate has been characterised by a doubling period of ten years, see table 1.4.

If this projection is correct, and if the 'energy gap' of the future is to be filled with nuclear power made available to the consumer in the form of electricity then much of the industrialised West will have gone a long way towards

Table 1.4 Electricity production, in units of 10^9 kWh$_e$, during recent years. Countries with annual production of less than 12×10^9 kWh$_e$ are not listed [Reproduced with permission from *Chemical and Engineering News*, **51**, 63 (1973)]

Country	1969	1970	1971	Average annual growth 1966–1971, %
North America				
Canada	190	204	215	6.3
USA	1553	1638	1718	6.6
Western Europe				
Belgium	29	31	33	7.7
France	132	141	148	6.9
FRG	226	243	260	7.9
Italy	110	117	122	6.3
Netherlands	37	41	45	10.3
Austria	26	30	29	3.9
Denmark	16	18	17	14.0
Finland	20	23	23	8.1
Norway	57	58	63	5.3
Spain	52	56	60	9.7
Sweden	58	61	66	5.6
Switzerland	30	32	33	3.2
UK	239	248	251	4.5
Eastern Europe				
Bulgaria	17	20	21	12.3
Czechoslovakia	43	45	47	5.3
East Germany	65	68	69	4.1
Hungary	14	15	15	4.8
Poland	60	65	70	8.1
Rumania	32	35	39	13.6
USSR	689	740	800	8.0
Yugoslavia	23	26	29	11.4
Latin America				
Argentina	15	17	19	9.9
Brazil	42	45	48	8.1
Mexico	26	29	31	10.4
Australasia				
Australia	51	56	60	8.4
New Zealand	13	14	15	5.3
Other				
India	49	55	59	11.5
Japan	316	359	379	12.0
South Africa	44	49	53	8.7

becoming an 'all-electric economy'. But is the widespread use of the synthetic fuel electricity desirable? Electric transmission is one of the most expensive methods of energy transportation. Direct comparisons of energy transport costs are very difficult because of varying end uses and the lack of parallel data. But if the cost of electrical tansmission by overhead lines is divided by three to compensate for the average 35% efficiency of conversion of fossil fuels to electricity, then electrical transmission is slightly more expensive than shipment of coal by conventional trains. This cost, is still about 70% higher than the projected costs of shipping coal by pipeline, or the cost of transporting natural gas by pipeline. These costs, in turn, are about three times higher than the cost of pumping oil through a pipeline, and several times higher than that of transporting oil in large supertankers (see figure 1.9).

Overhead transmission lines require about 12 acres per mile, and there is growing concern about the aesthetic effect of overhead lines. Underground transmission seems an acceptable alternative, but suffers from the disadvantage that the costs involved are at least nine (and sometimes up to 20) times as much as for overhead lines, for the same power carrying capacity.

Since it is almost impossible at the present time to store anything but trivial amounts of electricity this thus means that both the generating and transmission facilities have to be sized to cope with the maximum demand rather than the average demand, resulting in an average power plant load fraction of about 0.5. Although pumped hydro-electric storage appears economically

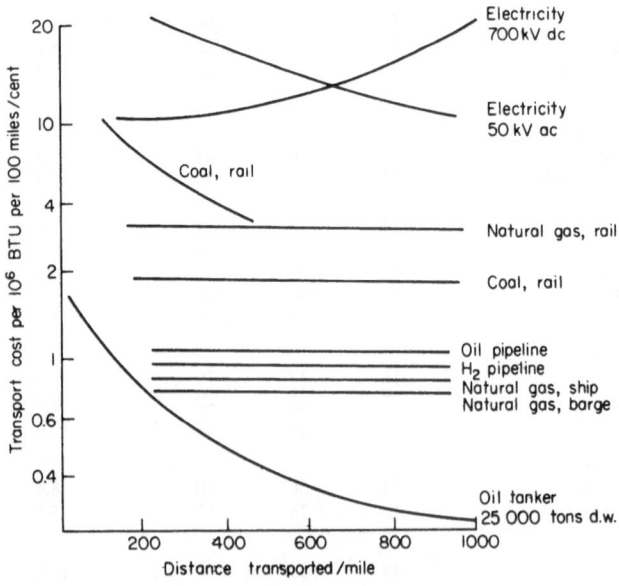

Figure 1.9 Transport costs of fuels. (Adapted from reference 5 with permission)

promising, suitable sites are strictly limited in number and capacity.

To provide a bridge between the limitations of supply and demand, it seems desirable, in the long term, to manufacture a synthetic fuel from non-fossil sources. The primary criterion for such a fuel is that, given unlimited energy, it must be available in unlimited quantity. It must therefore be synthesised from abundant materials, which in the long term excludes fossil fuels. Secondly it should be capable of being burned without the production of a noxious effluent, and thirdly the products of combustion must be assimilated into the environment at the point of use, without the need for recycling to the generating station. To meet these criteria only components of the atmosphere and oceans as sources can be considered as feasible alternatives in order to carry out the return cycle. Discounting any carbon-containing fuel (because of the extreme low concentration of atmospheric CO_2) and any nitrogen-containing fuel (because of the high concentration of noxious NO_x produced on combustion) the only attractive alternative is hydrogen.

In many respects hydrogen is the *ideal* fuel. Although it is not a 'natural' fuel, it can be readily synthesised from coal and natural gas. More importantly, it can be produced simply by splitting molecules of water with an input of electrical energy derived from a nuclear reactor. Perhaps the greatest advantage of hydrogen fuel, however, at least from an environmental standpoint, is the fact that when hydrogen burns its only combustion product is water.

In the hydrogen cycle the water produced rapidly equilibrates with the abundant and mobile water supply on the earth's crust. The cycle is characterised by negligible delay and does not disturb the environment, yet it relies on the environment to carry out the 'return' function. Our present system allows the combustion products of fossil fuels to remain in the immediate environment since the rate of formation of such products vastly exceeds the rate at which reconversion can take place. (This, basically, is the reason why the world is running out of fossil fuel reserves.) Assuming the availability of an abundant supply of nuclear or solar energy, the hydrogen system can be operated indefinitely and as rapidly as the demand requires without depleting any natural resources.

Few problems arise in considering the uses of hydrogen. It can be substituted for petroleum and coal in almost all industrial processes and furthermore its reducing properties make it an attractive component for use in the steel industry and for other metallurgical operations. The use of hydrogen as fuel would allow the industrial establishment to retain its present structure and would cause the least economic burden in the process of changeover from essentially a hydrocarbon based economy to one based on nuclear or solar energy. Moreover, a system which used large amounts of hydrogen would necessarily have at its disposal massive amounts of oxygen (the third largest product of the existing chemical industry) for which there are numerous demands.

Historically, hydrogen has long been recognised as an attractive fuel, since it is clean burning and is available from a renewable and universal raw material, water. It is not commonly used as a fuel because of difficulty in transporting the fluid. However, hydrogen is an essential ingredient in the chemical and oil industries, particularly in the manufacture of ammonia and in petroleum refining operations (see table 5.1). It is presently produced from fossil fuels, predominantly natural gas and petroleum, but since it is to supplant these materials future production must be from water. However, splitting water into its elements is an endothermic process; for example, even with advanced water electrolysis technology, it requires 1.3 BTU of electricity per BTU of heat from the H_2. Therefore, the contribution hydrogen may make in energy conservation is dependent on considerations of overall energy systems analysis including energy transport and ultimate end uses. The role of hydrogen in resource conservation is dependent on its substitutability for other fuels, which, in addition to preserving the finite fossil fuels for chemical use, can be a significant factor in providing a solution for the balance of payment problems facing the industrialised West. Interwoven with these considerations is the environmental impact of fuel production and use, which is of increasing concern and may have considerable effect on overall energy use, efficiency, and economics.

'Hydrogen economy' is a concept in which hydrogen, derived from nuclear or other energy forms, is used to supply all the demands commonly met today by fossil fuels, including industrial, commercial, residential, and vehicle power, as well as for the generation of electricity (figure 1.10). If such an idea appears, at

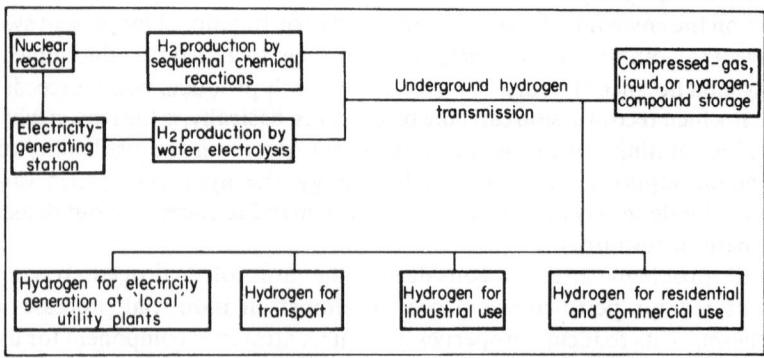

Figure 1.10 A hydrogen-economy design, based on nuclear energy. Solar energy as a source would require a modified diagram

first sight, revolutionary then a glance at figure 1.3 will remind us of a number of such energy revolutions during the past 100 years. In the future, new sources of power will require an efficient energy storage and transmission system, and whether or not electricity can fulfil such a role is questionable.

An important factor in implementing a hydrogen energy system is, of course, the cost of its production and use, and such costs must be compared to the costs of alternative fuels, and technological breakthroughs in hydrogen production will serve to speed up the introduction of the hydrogen system. Apart from the electrolysis route to hydrogen an alternative exists in using nuclear waste heat to decompose the water via a series of closed-cycle thermochemical processes. A system for producing hydrogen in an interrupted photosynthesis process has been proposed, as well as one using photons, produced from a fusion reaction, to split water. A more detailed discussion of hydrogen production will follow later.

As far as transmission is concerned the technology for moving a fluid via an underground pipeline is well established and is evident in the thousands of miles of natural-gas pipelines crisis-crossing many areas of the world. If the nuclear age comes stations will increase in size to reflect economies of scale and cooling water requirements will become the most important factor in siting. This factor, coupled with growing fears of plant safety, will require energy to be transported over longer distances than in the past. Hydrogen transmission via underground pipeline appears competitive with over-ground transmission of electricity at distances exceeding 300 miles (figure 1.11).

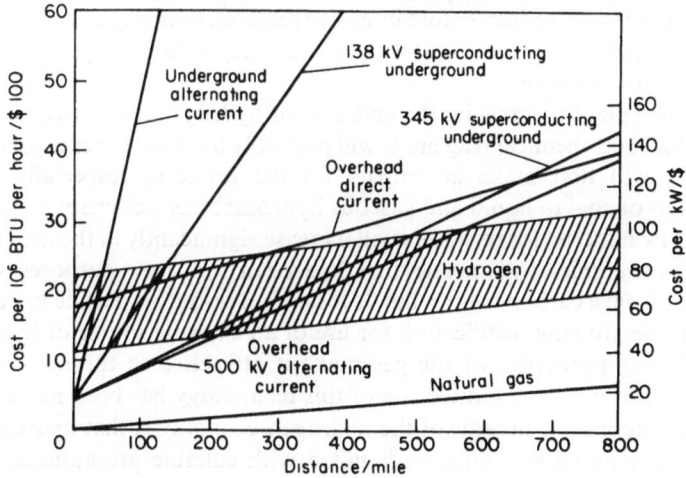

Figure 1.11 Cost of energy transmission facilities. (Reproduced from reference 6 with permission)

Although about three times the volume of hydrogen is required as of natural gas to deliver an equivalent amount of energy, hydrogen's density and viscosity are so much lower than those of methane that the same pipe can handle three times the flow rate of hydrogen.

In the hydrogen economy it will be possible to store vast quantities of hydrogen to even out daily and seasonal variations in load. At present, natural gas is stored in underground gas fields or in tanks as a cryogenic liquid. Whether gaseous hydrogen can be stored in underground porous-rock can only be ascertained by future field trials. However, 30 billion cubic feet of helium (a gas with similar leakage characteristics as hydrogen) is stored quite satisfactorily in an underground reservoir near Amarillo, Texas. If gaseous hydrogen is the preferred working medium, evaluation of gaseous storage is required. Mined caverns, aquifers, or depleted gas wells may be capable of use for large-scale storage.

Liquid hydrogen storage has been practised in volumes up to 1 000 000 gallons. The technology for liquefaction and tankage has already been developed, mainly for the space industry. In the gas industry, cryogenic storage of natural gas for peak-shaving purposes is already practical. Whilst the costs involved in H_2 storage are far higher than corresponding costs for natural gas, with further development and use of mass production, it is expected that present costs could decrease by as much as a factor of two.

For hydrogen to be used as fuel for transport, a portable storage system is required. In fact, efficient storage appears to be a far greater problem than modification of power systems to operate on hydrogen fuel. Storage of hydrogen via metal hydrides at densities near that of liquid hydrogen, offers the best prospect of developing a fuel tank. Probably, this aspect of the hydrogen economy requires more intensive research than any other, but unless a very inexpensive system is developed this may inhibit large-scale use of hydrogen in transport.

It is convenient to consider the end-uses of hydrogen under two headings: namely fuel and chemical reagent. It will probably be in the latter capacity that hydrogen will first make an impact on the economy, especially in the conversion of coal to liquid and gaseous hydrocarbons (see chapter 7) and its use as a raw material for fertilisers will increase significantly in the near future. Initially, economics will dictate that hydrogen for these purposes will be produced from a carbon source, but as fossil-fuel resources decline in quantity there will be growing justification for use of an external source of hydrogen. The reducing properties of the gas make it attractive to the metallurgical industry, although only limited use of this technology has been made so far.

One of the main criticisms of the hydrogen-economy is that hydrogen is a hazardous material and must be handled with suitable precautions. If it is handled properly, however, in equipment designed to ensure its safety, it is no more dangerous than existing fuels. It is the attitude adopted by the public at large which is the important factor. For example, people were quite happy to use town gas for domestic use—whilst few realised that it contained up to 80 % hydrogen. Nowadays motorists are relatively unconcerned about travelling at high speeds along crowded motorways carrying tankfuls of petrol which are potential fire-bombs. Very strict codes of conduct are enforced today for use of hydrogen in industry and there seems no reason why such codes could not be implemented on a more general scale.

The above, then, is a scenario for the hydrogen economy. The change-over need not be a drastic one, since many appliances are well suited to run on both hydrogen or conventional fuels. How soon such an economy can be introduced will depend on many factors—not least of which are the rate of price increase of conventional fuels and the rate of development of nuclear or solar energy.

CHAPTER 2

PRODUCTION OF HYDROGEN

Perhaps the most important factor influencing the viability of an energy system based on a synthetic fuel is the fuel's production. At present, almost all the hydrogen produced worldwide is derived from hydrocarbon sources. But, as pointed out previously, since hydrogen is to replace fossil-fuels in the long term, alternative raw materials will have to be found and alternative processes developed to produce hydrogen. The only 'endless' source of hydrogen is the sea. At present electrolysis of water is limited to areas where cheap hydro-electric energy is available. As oil and gas, the traditional sources of hydrogen, become too expensive to use in this context, water will gradually take over as the sole raw material, and in the long term it is likely that alternative energy sources will provide the necessary heat.

Hydrogen from fossil fuels

At present the raw materials used to produce hydrogen are natural gas, coal, oil, and steam. In some processes the combustion of the fuel simply provides the energy necessary to release the hydrogen from water, i.e. water-gas generation. In others, such as the 'steam reforming' of natural gas, the hydrogen from water is supplemented by the hydrogen in the fuel, whilst in some, all the hydrogen is derived from the fuel and no water is used (i.e. partial combustion).

Availability and cost of hydrocarbon dictates the process to be employed. Thus prior to 1939 about 90% of the hydrogen produced in the USA was by the water-gas reaction; nowadays the principle processes are variations of steam reforming of natural gas. So far, the extreme low price of natural gas has enabled it to compete successfully with coal, even in areas where the latter is readily available.

Steam-reforming is at present the most important process for hydrogen manufacture. Natural gas and steam are passed over a fired-bed catalyst, usually nickel-based, at temperatures of 650–700 °C to produce mixtures of hydrogen and carbon monoxide. Because of nickel's sensitivity to sulphur, desulphurisation of the feed gases is usually necessary, and in cases where refinery gas is used it is necessary to remove unsaturated hydrocarbons to prevent coking of the catalyst. Operating pressures of up to 10^3 atmospheres are used. In a further reaction vessel additional steam is added in order that unwanted carbon monoxide can react to give more hydrogen and also carbon dioxide (the shift reaction). The latter is usually removed by scrubbing with either water or aqueous carbonate solution. The overall reaction can be written:

$$C_nH_{2n+2} + 2nH_2O \rightarrow nCO_2 + (3n+1)H_2$$

An alternative to steam reforming is the reforming of *petroleum naphtha* under pressure. The purpose of such an operation in a refinery is to upgrade the octane rating of gasoline. Cyclisation and dehydrogenation of the straight-chain feedstock are the principle reactions involved, and large quantities of by-product hydrogen are produced. However, no hydrogen leaves the refinery as such, since it serves a valuable purpose in hydrocracking fuel oils to give lighter fractions. Lower proportions of by-product hydrogen are produced by other refinery operations, such as catalytic cracking of gas oil or thermal cracking of naphtha to ethylene.

For those hydrocarbons with very high carbon/hydrogen ratios and which also contain a high percentage of sulphur, *partial oxidation* provides a convenient route for hydrogen manufacture, and essentially involves combustion of the organic compound in an inadequate supply of oxygen; heavy fuel oil usually needs additional steam. Optimum temperatures are in the range 1300–1500 °C and hence careful control of feed-rates into the combustion chamber is required. The product gas, consisting almost entirely of hydrogen and carbon monoxide, is desulphurised and subjected to 'shift conversion'. Overall, the reaction can be represented by:

$$2C_nH_m + nO_2 + 2nH_2O \rightarrow 2nCO_2 + (2n+m)H_2.$$

The process of *carbonisation* is the oldest method for making hydrogen, and involves heating coal to about 900 °C in the absence of air. This results in direct decomposition of the coal into liquid, gaseous, and solid products. The actual nature of these products is, of course, important to the overall economics of the process. Carbonisation is operated to make coke for use in steel manufacture, in 'coke ovens', or to make gas (i.e. town gas) in retorts.

As stated earlier, where gaseous or liquid hydrocarbons are available then these materials are the ones preferred for hydrogen manufacture. Vapour phase reaction leads to a very much simpler and cheaper plant, which requires less labour and is much more thermally efficient than a coal-based plant. For example, the thermal efficiency for the steam reforming process is about 72%, whilst that of coal gasification is around 53%, and this is reflected in overall production costs. Pre-1973 costs for hydrogen via coal gasification were about \$1.30/10^6 BTU, whilst those for hydrogen via steam reforming were 80 cent/10^6 BTU.

New production methods for hydrogen
Assuming the premise that, in the long term, only water will provide an

economic source of hydrogen, then the critical factor to be considered is how more novel energy sources will be used in the 'cracking' process. New sources of energy are assumed to be nuclear (fission and fusion), solar (direct), geothermal and wind, and the processes available are:

(1) electrolytic
(2) thermo-chemical
(3) biological
(4) radiolytic

Direct thermal splitting of water requires temperatures in excess of 2800 °C, and only fusion reactors might be expected to make available such temperatures. Although some schemes have been proposed, it appears to be an impractical process for hydrogen production. Only the biological method is limited to one source of energy, and choice of primary energy source exists for all other processes. Of the processes listed above, only electrolysis is a mature technology—commercial plants having been in existence for some considerable time.

It is now proposed to examine the potential of each process, both from a technical and economic viewpoint.

Physical principles and theory of electrolytic hydrogen production

Water electrolysis is accomplished by passing a direct current between two electrodes immersed in an electrolyte (usually potassium hydroxide solution); hydrogen forms at the cathode and oxygen at the anode. The amount of hydrogen produced is directly proportional to the current passing between the electrodes and is given by Faraday's law as one pound hydrogen and eight pounds oxygen per 12 000 amp-hours of electricity.

The energy which must be supplied to the cell to cause the reaction H_2O (liquid) $\rightarrow H_2$ (gas) $+ \frac{1}{2}O_2$ (gas) to proceed is the enthalpy of formation of water, ΔH, and is equal to 68.32 kcal/mol at 25 °C and one atmosphere pressure. However, only the free energy of this reaction, ΔG, equal to 56.69 kcal/mol, has to be supplied to the electrodes as electrical energy; the remainder is required as heat, and this can theoretically be provided as thermal energy from the surroundings, or from electrical losses within the cell.

According to a fundamental law of thermodynamics, the electrical work, ΔG, done on or by a cell is equal to the free energy change occurring, or

$$\Delta G = -nFE$$

where n is the number of electrons transferred during electrolysis, E is the reversible voltage of the cell, and F is Faraday's constant. By use of this law, the minimum theoretical electrical energy requirement can be measured in terms of applied voltage, and for the electrolysis of liquid water solutions at 25 °C it is 1.229 V, or 14.9 kWh/(lb hydrogen). A perfect cell would operate at this voltage and energy input, but would require the additional input of thermal energy equivalent to another 3.1 kWh/(lb hydrogen). In order to provide all the necessary energy as electrical energy, the corresponding voltage is 1.484 V (18.0 kWh/lb). A practical cell can approximate to this voltage at low output rates, since it is still experiencing a 20 % loss of efficiency from an 'ideal' situation. Under usual operating conditions, commercial electrolysis plants require much higher power levels, due to even greater than 20 % power losses in electrolyte or in the electrodes themselves. The theoretical reversible voltage (defined by the free energy change) decreases with temperature as shown (figure 2.1). Raising the electrolyte temperature lowers the voltage at which

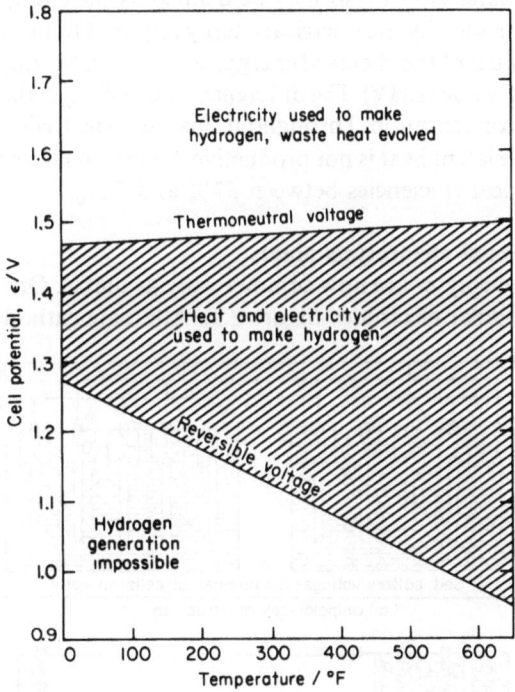

Figure 2.1 Idealised operating conditions for water electrolysis. (Reproduced from reference 6 with permission)

water can be decomposed. Moreover, higher temperatures mean that electrode processes occur at faster rates with lower losses; on the other hand, the entropy change for the formation of water is negative, which means that in the case of fuel cells the useful energy output falls with increasing temperature. The voltage corresponding to the enthalpy change, also called the 'thermoneutral' change, increases only slightly with increasing temperature (figure 2.1). There are three parts to this figure:

  (i) no hydrogen is evolved

  (ii) hydrogen is made at greater than 100 % electrical efficiency

  (iii) hydrogen is made at less than 100 % efficiency with the production of waste heat

In actual fact an electrolyser does not operate at all at the theoretical 'reversible voltage' since here the rates of the electrode processes are zero and for water decomposition, a higher voltage must be applied. This excess voltage, or 'overvoltage', is related to the current that passes through the cell. Higher currents per unit area of electrode require higher overvoltages and hence lower efficiencies. Overvoltages are reduced by increasing the operating temperature, by proper design of electrodes, and by the incorporation of catalysts into the electrodes.

Efficiency of water electrolysis may be defined as the energy stored in the hydrogen (ΔH) divided by the electrical energy required to produce hydrogen. There are two values of the chemical energy, i.e. the 'high heating' value (HHV) and the 'low heat' value (LHV). The difference, about 20%, is the heat available as latent heat of condensation. Throughout this work the LHV is used, since in most end uses the latent heat is not productive. Commercial electrolysis plants operate at electrical efficiencies between 57% and 72%.

Current commercial electrolysis plants

Two types of electrolyser are now commercially available. One employs tank cells with mono-polar electrodes (figure 2.2). Alternate cathodes and anodes

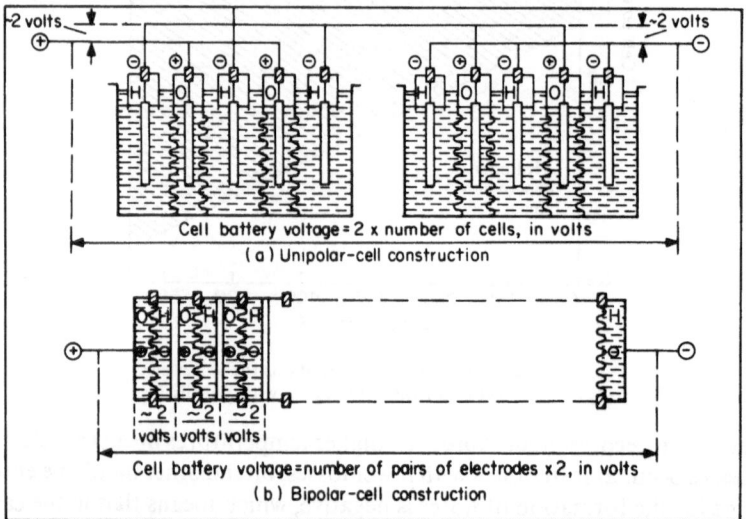

Figure 2.2 Schematic diagram showing unipolar and bipolar cell construction. O and H represent oxygen and hydrogen evolution. (Reproduced from reference 6 with permission)

are separated by power diaphragms to prevent gas mixing. A large iron vat holds the alkali electrolyte, and electrodes of alternative polarity are suspended in the vat. These electrodes consist of flat sheets of mild steel welded to steel bus bars. Plates are heavily nickel-plated, but the cathodes are left plain.

Usually, a bag or skirt of asbestos is used as a diaphragm to maintain separation of the gases. Each tank operates as one cell and connecting tanks in parallel keeps the required voltage down to two volts and permits high current densities.

The alternative to tank-type cells is the so-called filter-press design which contains alternate layers of electrode and diaphragms. The electrodes in this arrangement are solid metal and bipolar, i.e. cathodic on one side and anodic on the other, and each electrode spans the entire cell cross-section. Only end electrodes are connected to external power sources: the individual cells being connected in series. The advantages of such an arrangement are modular design, smaller space requirements and higher operating pressures.

The electrolyte most commonly used, in both types of design, is potassium hydroxide solution at a strength of 28 % for maximum conductivity. Aqueous acid electrolyte usually causes corrosion of cell materials. At above ambient temperatures, higher strengths of hydroxide are used to reduce loss of feed water by evaporation. In the cases where electrolyte is in contact with air sodium hydroxide is used for, although it exhibits lower conductivity, it costs less to replace. However, the original electrolyte charge is normally used for more than ten years in sealed cells, though periodic topping-up is required to replace mechanical losses.

Installed electrolysis plant capacity throughout the world was estimated to be 3×10^2 pounds of hydrogen per day in 1972, and this is mainly used in the manufacture of ammonia for fertiliser. The low cost of producing hydrogen from fossil fuel means that hydrogen production via electrolysis only accounted for some 3 % of the total amount of hydrogen consumed in the USA. Such electrolysis plants are located in areas where there is significant demand for fertiliser, plentiful supply of low-cost electricity, and a limited amount of hydrocarbon fuel. Areas where the above conditions prevail include Egypt, Chile, India, and Norway.

Table 2.1 lists a number of major hydrogen plants, their operating parameters and capacities. It is significant that a large number of plants have been operating for a considerable number of years. Figure 2.3 shows the efficiency of such plants in terms of amount of energy required to produce one pound of hydrogen. An electrical efficiency of 100 % is achieved if performance reaches the $\Delta H(g)$ line illustrated, assuming the LHV of the hydrocarbon is adopted.

Economics of hydrogen production
In the design of an electrolysis plant two factors predominate in the determination of the cost of the hydrogen produced: (1) the capital cost of the electrolysis plant suitably discounted and (2) the operating cost—essentially the electric power cost.

Capital cost is very dependent upon the size and nature of the electrodes, and thus low capital cost tends to push the design operating current density to the highest possible level. But this higher power consumption results in lower

Table 2.1 Summary of Electrolytic plant equipment
(Prepared by Teledyne Isotopes, Inc.)

Company, location	Cell name	Type	Current density/ amp ft^{-2}	Emf/ V cell^{-1}	Module size lb H$_2$ day^{-1}	Pressure psi	No. of Plants	Largest size lb H$_2$ day	Earliest plant year	Best known plant year
A. Norsk Hydro Notodden, Norway	Hydro-Pechkranz	Filter press	140	1.773	1800	1	3	284 000	1927	Rjakon, Norway 1965
B. Lurgi Frankfurt, Germany	Zdansky-Lonza	filter press	200	1.832	4200	440	32	22 000	1955[a]	Cuzco, Peru 1958
C. DeNora Italy	DeNora	filter press	280	2.00[b]	4100	1	2	110 600	1958	Nangal, India 1958
D. Pintsch-Bamag Germany	Bamag	filter press	230	1.788	2600	13.5	200	UNK	1935	UNK
E. Electrolyzer Corp. Canada	Stuart	tank	200	2.04	40	0.03	1000	1130	1930	Teledyne Wah Chang, Alabama, USA 1971
F. Cominco Canada	Trail	tank	80	2.142	38	0.1	1	77 000	1939	Trail Canada 1939
G. Teledyne Isotopes USA	EGGS	filter press	400	2.1	65	70	2	50	1968	Teledyne Isotopes USA 1972
H. Demag Elektro-metallurgic GMBH Duisburg, Germany	Demag	filter press	92 to 280	1.75 to 1.95	900	1.0	57	177 000	1945	Aswan Dam Egypt 1960
I. Electric Heating Equipment Co. USA	Kent	tank	115	2.2	28	0.1	100	1610	1920	Hobart, Tasmania 1949
Cells being developed										
J. Teledyne-Isotopes USA		filter press	400	1.65	13	2000		Designed for military aircraft application		
K. Teledyne Isotopes		filter press	250	1.64	94	3000		Designed for nuclear submarine application		
L. General Electric		solid electrolyte	3260	1.2 to 1.8		1		2000 °C not now under development		
M. Westinghouse		solid electrolyte						Used for CO$_2$ electrolysis in spacecraft atmosphere control system.		

[a] First Zdansky-Lonza plant   [b] DeNora has indicated an ability to achieve 1.61 V on new cells.

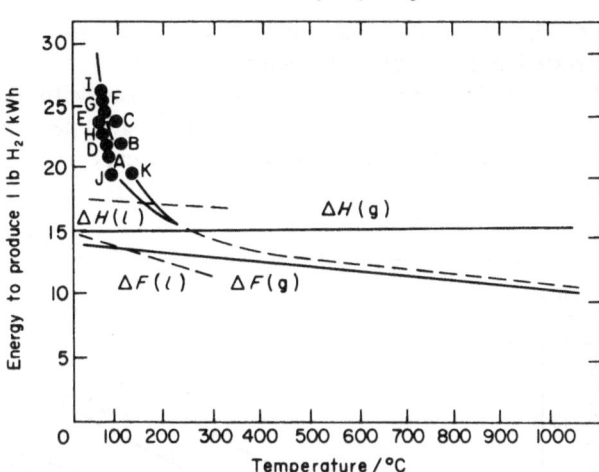

Figure 2.3 The efficiency of various hydrogen producing plants: for
the significance of A–K see table 2.1

operating efficiency. Increased current density may also be obtained by using precious metal catalysts or extremely complicated electrode construction. Some degree of optimisation is usually employed and the actual operating parameters vary according to the cell's application and the cost of power. In 1972 capital costs of large-scale electrolysis plants were \$95/(lb hydrogen) per day, or approximately \$95/kW based on input power of 24 kWh/(lb hydrogen).

The major operating cost is the price of electricity. Even in the case of off-peak electricity, say at 0.3 cents/kWh (1970 US average price), power costs can still represent as much as 75 % of the overall cost.

Thus overall cost of hydrogen production can be represented by an empirical equation of the following type:

$$C_T = aC_0 + b$$

where C_T is the total production cost of hydrogen (in \$/10^6 BTU), C_0 is the cost of electricity (in tenths of a cent/kWh), b represents a fixed cost factor of capital equipment depreciated at a given factor per year, a is an energy conversion factor.

In the design of an electrolyser, factors a and b are variable but C_0 is fixed. Thus, desirable features of any plant are low unit costs, long life-time, and high utilisation.

Factor a has the greatest effect on overall costs. It represents the efficiency with which electrical energy can be transformed into chemical energy stored in the hydrogen. Although ac–dc rectification efficiency and cell current efficiency modify overall performance, the major loss of energy is that required to overcome ohmic resistance of electrolyte as well as anodic and cathodic overpotentials. The value of a becomes 0.294 (due to units employed) if energy is converted at 100% overall efficiency.

Figure 2.4 shows the variation of production costs with the price of electricity. This assumes capital costs of \$95/(lb hydrogen) per day and fixed charge rate of 15% and a 90% plant factor; an electrical efficiency of 60% is also assumed.

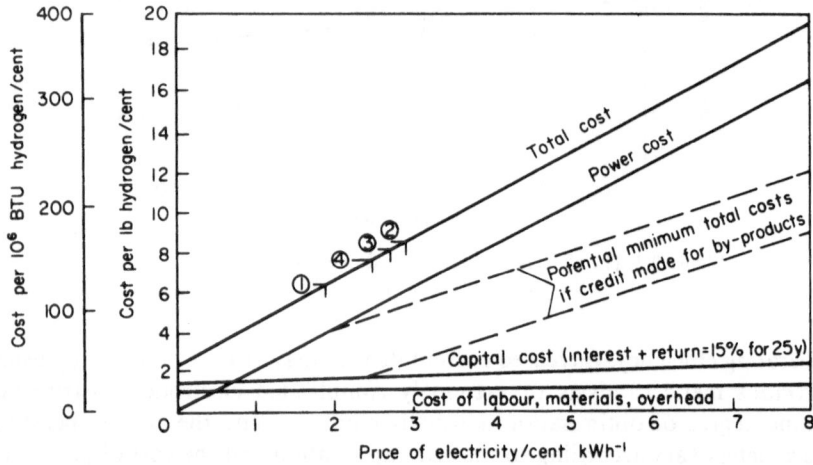

Figure 2.4 Total cost of hydrogen production: 1, fuel oil at \$2.5 per bl; 2, naphtha at \$30 per ton; 3, coal gasification at \$7.50 per ton of coal; 4, natural gas at \$0.75 per 10^6 BTU

The future for electrolyser systems

Although water electrolysis is already a relatively efficient process, further improvements are inevitable and desirable. Existing large-scale electrolyser plants are all operated at current densities of about 100–200 A/ft^2 and at voltages of about 2.0 ± 0.1 V. Increasing the operating temperature and reducing some of the internal ohmic losses appear possible, and a reduction of 25–35% in power requirement is certainly possible. With an electricity-intensive process such as electrolysis, considerable leverage exists for decreasing power needs per unit of production or in decreasing the cost of power.

The solid polymer electrolyte

Although the electrolytic manufacture of hydrogen has been essentially a European technology, in the USA, space and military applications have recently caused an increase in interest in electrolyser technology, leading to a number of advanced concepts for electrolyser design and construction which promise to reduce the cost of hydrogen manufacture.

The most important development in electrolyser design is the recent introduction of solid polymer as the electrolyte. In comparison with aqueous electrolyte systems, this allows use of severer conditions of temperature and pressure with minimum maintenance costs and, because the electrolyte is solid, the catalytic electrodes are not required to retain the electrolyte and therefore can be optimised for catalytic activity at minimum costs. The solid polymer electrolyte (SPE) is a solid plastic sheet of perfluorinated sulphonic acid polymer similar to Teflon, which when saturated with water is a very good electrical conductor (< 15 ohm cm resistivity). This is the only electrolyte required and there are no free acid or alkali liquids in the system. Ionic conductivity is provided by the mobility of the hydrated hydrogen ions (H^+, H_2O). These ions move by passing from one sulphonic acid group to another. The sulphonic acid groups are fixed, keeping the acid concentration within the electrolyte constant. A thin layer of platinum black (1 to 5 mg/cm^2) is attached to the polymer surface to form the hydrogen electrode. A similar layer of alloy catalyst forms the oxygen electrode.

In the USA the General Electric Company is putting a lot of effort into developing an electrolysis process based on a solid polymer electrolyte in a programme sponsored by General Electric, the US Department of Energy, and some local electric power producers. The goal of the programme is to achieve an overall efficiency of more than 85 % at a capital cost of less than $100 per kW, with electrolyser cell lives of 40 000 hours or more and a total system life of more than 20 years. In early 1978 they announced that if all these factors could be accounted for a scale-up to a 5 MW demonstration unit would follow. Hydrogen from such a system is estimated to cost $5.00 per million BTU, without a credit for by-product oxygen. In 1977 the Illinois Gas Institute estimated that a nuclear power plant could produce electrolytic hydrogen at $5.36 per million BTU without oxygen credit ($4.73 with oxygen credit). Thus the General Electric scheme looks very attractive thus far. They have lately achieved cost reductions through development of moulded carbon and phenolic separator/current collectors which replace original transition-metal screens. Better gasket materials have also been developed which permit operation at higher gas pressures (up to 600 psi).

Thermochemical water splitting
Using direct heat to crack water is attractive because of its efficient use of energy, since if hydrogen is produced via electrolysis thermal energy has first to be converted into mechanical energy to generate the electricity, and the efficiency of such a step is thermodynamically limited. Highest temperatures achieved in the electrolysis plant are limited by the nature of the materials employed, and hence maximum efficiency is no greater than 40 %. This figure thus sets the upper limit for the overall efficiency possible by an electrolytic route. By-passing this low conversion stage and using the thermal energy directly is obviously attractive. At temperatures available from fission reactors, or even at those envisaged from fusion processes, only about 1 % of the water is dissociated in a conventional single-step reaction. Table 2.2 gives the

Table 2.2 Thermal splitting equilibrium data for water

Pressure/atm	Temperature/K	K_a	Mole fraction of H
0.1	1000	6.75×10^{-10}	2.09×10^{-6}
0.1	2000	3.57×10^{-3}	7.39×10^{-2}
0.1	3000	6.84×10^{-1}	6.67×10^{-1}
0.1	4000	9.53	6.67×10^{-1}
1.0	1000	6.75×10^{-10}	1.21×10^{-6}
1.0	2000	3.57×10^{-3}	3.58×10^{-2}
1.0	3000	6.84×10^{-1}	6.46×10^{-1}
1.0	4000	9.53	6.67×10^{-1}

equilibrium constant, conversion, and mole fraction of hydrogen produced as a function of pressure and temperature.

It is not difficult to see why temperatures of about 2500 ° C are required for a reasonable yield of hydrogen. The energy required for the endothermic reaction is the enthalpy change and is composed of two parts: a thermal energy requirement, ΔQ, and a useful work requirement, ΔW. Under conditions of reversibility and at constant temperature and pressure the following conditions hold:

$$\Delta W = -\Delta G$$

$$T \Delta S = \Delta Q$$

where ΔG and ΔS represent changes in Gibb's free energy and reaction entropy respectively. The rate of change of free energy with increasing temperature is also important and is represented by:

$$\left[\frac{\partial \Delta G}{\partial T} \right]_P = -\Delta S$$

It is clear that a lowering of the work requirement in conjunction with a corresponding increased thermal energy requirement will yield more efficient use of the available heat. The low thermal efficiency (defined as the ratio of the heating value of the hydrogen to the total thermal energy input) of direct cracking results from a slow decrease in the work requirement as the temperature is increased, i.e. the entropy change is not large enough. But it is now possible to devise a sequential closed-cycle series of chemical reactions in which hydrogen and oxygen are produced, water is consumed, and all chemical products are recycled—and in such a system the entropy change is not constant and may vary depending upon the reaction employed. This method offers the possibility for chemical production of hydrogen and oxygen from water using direct nuclear heat.

In the enthalpy against temperature diagram (figure 2.5) a two-step

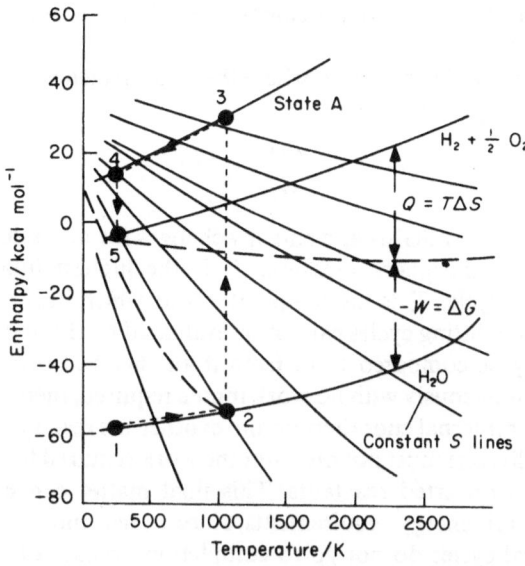

Figure 2.5 Enthalpy against temperature diagram for
water decomposition

thermochemical cycle for cracking water is nicely illustrated. Solid lines marked H_2O and $H_2 + \frac{1}{2}O_2$ represent the thermodynamic state of water vapour and a stoichiometric mixture of hydrogen and oxygen resulting from the water decomposition. The distance between these two state lines is the enthalpy change in going from the initial state of water to the final state at a particular temperature. This enthalpy change may be brought about by the combination of heat ($T \Delta S$) and work ($-\Delta G$) in the proportions indicated by the broken line ($Q-W$) running between the two state lines. Any chemical reaction can be characterised by these three lines in the H against T diagram.

Water vapour dissociation at room temperature involves an enthalpy change of 57.8 kcal mol^{-1} between states 1 and 5. But in principle it is possible to go through an intermediate state A with only the addition of heat (path 1–2–3–4–5) in a two reaction system. This can be done by carrying out the fast reaction in the cycle at a temperature above the intersection of its $Q-W$ line with the reactants enthalpy curve. So far, however, there has been little success in finding such an intermediate state A. Oxides and hydrides recently investigated exhibit $Q-W$ lines which result in zero work requirement only at unreasonably high temperatures.[7]

However, for cycles involving more than two reaction steps the possibilities of minimising work requirements are greatly increased. Similar thermodynamic conditions apply to every reaction in the sequence and the net enthalpy, entropy, and free energy changes must equal those of the one-step water decomposition. Net work requirements are different, since work is not a function of state.

A relationship has been derived where the ideal efficiency of any cycle is only a function of temperature, i.e.

$$\leqslant \frac{\Delta H}{\Delta G} \frac{T_r - T_e}{T_r}$$

ΔH and ΔG refer to enthalpy and entropy changes for water decomposition at 25 °C and one atmosphere pressure, T_r is the highest input temperature available to the cycle, T_e is the temperatures at which heat is rejected.

Actual water splitting cycles cannot be evaluated by the above relationship, but can merely be compared to it. Even if all chemical reactions of a cycle proceeded spontaneously with no work inputs required, there still remains the requirement for internal endothermic and exothermic reactions to be perfectly matched. Finally, one must not discount the work required for separation and recycling of unconverted reactants. This final matter can be a substantial burden on overall energy requirements, since in fact most reactions used in thermochemical cycles do not go to completion because of the thermodynamics involved. In most cases large amounts of unreacted starting materials are present with the products in the effluent of a reactor. These must be separated, the reactants recycled to the reactor, and the products circulated to appropriate steps of the process.

In selecting cycles which are practically acceptable the most important criterion required is that the free energy changes of individual reactions must approach zero within the temperature range considered. Large negative free energy changes characterise products that are relatively too stable to be recycled and subsequently reacted, and cycles containing such reactions cannot be closed efficiently. Temperatures must be selected so that each step is run under minimum energy requirements so that efficient heat exchange between individual steps is achieved. Conditions for minimum energy requirement are that those reactions with positive entropy changes should be run at high temperatures and those reactions with positive energy changes should be run at low temperatures.

Current status of technology of thermochemical processes

Thermochemical processes producing hydrogen are not in commercial use today but several have reached pilot plant stage. However, it must be admitted that most schemes are still at the laboratory stage and the main areas of research are presently concerned with the evaluation of thermodynamic and kinetic properties of possible cycle reaction steps.

During the last 5–10 years a number of groups around the world have searched for acceptable water-splitting cycles. The largest effort by far is at the Euratom Laboratory at Ispra, Italy, headed by Dr Marchetti. This project began in 1969 and is presently employing fifty people. Other active projects include German efforts at Julich (KFA) and the University of Aachen, and the USA's efforts are spearheaded by the Institute of Gas Technology, Argonne National Laboratory, Los Almos National Laboratory, General Electric Company, and General Atomic Company. About thirty cycles have been published in the literature and undoubtedly many more will soon appear. A representative few will be discussed here.

Some of the most promising cycles have been based on the chemistry of halide compounds. One of these processes using calcium bromide, and which unfortunately also involves mercury, was named 'Mark 1' by DeBeni in 1970[8]:

$$CaBr_2 + 2H_2O \rightarrow Ca(OH)_2 + 2HBr \qquad 730\,°C$$

$$Hg + 2HBr \rightarrow HgBr_2 + H_2 \qquad 250\,°C$$

$$HgBr_2 + Ca(OH)_2 \rightarrow CaBr_2 + HgO + H_2O \quad 200\,°C$$

$$HgO \rightarrow Hg + \tfrac{1}{2}O_2 \qquad 600\,°C$$

However, this cycle has attractive characteristics, which include: easy product separation and practically 100% recovery of chemicals for recycling. Moreover, the highest reaction temperature of 730 °C is well within the range of present high temperature gas cooled reactors. DeBeni and Marchetti[9] estimate efficiencies in the Mark 1 process to be 40–60%. The main disadvantages involve difficulties in handling hydrobromic acid and the contamination possibilities inherent in the use of mercury.

In every thermochemical cycle, at least one element is required which is able to change its oxidation state. It is well known that iron tends to a higher valence when it is an oxide than when it is a chloride. This, then, is the reasoning behind the use of these three elements iron, chlorine, and oxygen in many proposed cycles (one by Professor Knoche at Aachen University[10], four by Hardy[11,12], one at General Electric[13] and others elsewhere). Table 2.3 summarises the work.

The five-step sequence illustrated from Aachen University requires excessive recirculation and high temperatures, making this cycle unattractive. On the other hand, thermal efficiencies evaluated for the Mark 7 and Mark 9 processes appear to show more promise. Of the cycles proposed by General Electric, the 'Agnes' one is the only process published which uses iron as the transition element and, based on a limiting Carnot efficiency of 58%, an overall efficiency of 41% is thought possible. The use of other metals and transition elements with chlorine has also been proposed (table 2.4).[13-15] Of these, the Beulah cycle has the highest thermal efficiency reported (53%).

The only two step process published[16] consists of a reverse Deacon reaction followed by electrolytic decomposition of hydrogen chloride. The steps are outlined below:

$$H_2O + Cl_2 \rightarrow 2HCl + \tfrac{1}{2}O_2 \qquad 700\,°C$$

$$2HCl \rightarrow H_2 + Cl_2 \text{ (electrolysis)} \qquad 300\,°C$$

Hydrogen and Energy

Table 2.3 Cycles using iron, chlorine, and oxygen

Knoche, Aachen University

$3Fe + 4H_2O \rightarrow Fe_3O_4 + 4H_2$	500 C
$Fe_3O_4 + \frac{9}{2}Cl_2 \rightarrow 3FeCl_3 + 2O_2$	1000 °C
$3FeCl_3 \rightarrow 3FeCl_2 + \frac{3}{2}Cl_2$	350 °C
$3FeCl_2 + 3H_2 \rightarrow 3Fe + 6HCl$	1000 °C
$6HCl + \frac{3}{2}O_2 \rightarrow 3H_2O + 3Cl_2$	500 °C

Hardy, Euratom, Mark 9

$6FeCl_2 + 8H_2O \rightarrow 2Fe_3O_4 + 12HCl + 2H_2$	650 °C
$2Fe_3O_4 + 3Cl_2 + 12HCl \rightarrow 6FeCl_3 + 6H_2O + O_2$	175 °C
$6FeCl_3 \rightarrow 6FeCl_2 + 3Cl_2$	420 °C

Hardy, Euratom, Mark 7

$3H_2O + 3Cl_2 \rightarrow 6HCl + \frac{3}{2}O_2$	800 °C
$18HCl + 3Fe_2O_3 \rightarrow 6FeCl_3 + 9H_2O$	100 °C
$6FeCl_3 \rightarrow 6FeCl_2 + 3Cl_2$	400 °C
$6FeCl_2 + 8H_2O \rightarrow 2Fe_3O_4 + 12HCl + 2H_2$	600 °C
$2Fe_3O_4 + \frac{1}{2}O_2 \rightarrow 3Fe_2O_3$	400 °C

Wentorf and Hanneman, General Electric, 'Agnes'

$3FeCl_2 + 4H_2O \rightarrow Fe_3O_4 + 6HCl + H_2$	450–750 °C
$Fe_3O_4 + 8HCl \rightarrow FeCl_2 + 2FeCl_3 + 4H_2O$	100–110 °C
$2FeCl_3 \rightarrow 2FeCl_2 + Cl_2$	300 °C
$Cl_2 + Mg(OH)_2 \rightarrow MgCl_2 + \frac{1}{2}O_2 + H_2O$	50–90 °C
$MgCl_2 + 2H_2O \rightarrow Mg(OH)_2 + 2HCl$	350 °C

The thermodynamics of the hydrogen chloride – hydrogen – oxygen – water system have been widely studied.[17] At one atmosphere pressure and 730 °C, the reverse Deacon reaction proceeds with a 60 % conversion of water. Hence, such a step has been included in many other proposed cycles, e.g. the vanadium chloride process. Moreover, since electrolysis of hydrogen chloride requires less energy input than electrolysis of water further development of this cycle is likely.

Cycles based on the reaction of water with alkali metals to produce hydrogen and metal–oxygen compounds have also been proposed, and one[18] employing caesium is:

$$2H_2O + 2Cs \rightarrow 2CsOH + H_2 \qquad 100\ °C$$

$$2CsOH + \frac{3}{2}O_2 \rightarrow H_2O + 2CsO_2 \qquad 500\ °C$$

$$2CsO_2 \rightarrow Cs_2O + \frac{3}{2}O_2 \qquad 700\ °C$$

$$Cs_2O \rightarrow 2Cs + \frac{1}{2}O_2 \qquad 1200\ °C$$

Table 2.4 Cycles using other metals and chlorine

Funk and Reinstrom, Allison Div. General
Motors[14]

$H_2O + Cl_2 \rightarrow 2HCl + \frac{1}{2}O_2$	700 °C
$2VCl_2 + 2HCl \rightarrow 2VCl_3 + H_2$	25 °C
$4VCl_3 \rightarrow 2VCl_4 + 2VCl_2$	700 °C
$2VCl_4 \rightarrow 2VCl_3 + Cl_2$	25 °C

Knoche, Auchen University[15]

$H_2O + Cl_2 \rightarrow 2HCl + \frac{1}{2}O_2$	900 °C
$2HCl + 2CrCl_2 \rightarrow 2CrCl_3 + H_2$	200 °C
$2CrCl_3 \rightarrow 2CrCl_2 + Cl_2$	1000 °C

Wentorf and Hanneman, General Electric
'Beulah'[13]

$2Cu + 2HCl \rightarrow 2CuCl + H_2$	100 °C
$4CuCl \rightarrow 2CuCl_2 + 2Cu$	30–100 °C
$2CuCl_2 \rightarrow 2CuCl + Cl_2$	500–600 °C
$Cl_2 + Mg(OH)_2 \rightarrow MgCl_2 + H_2O$ $+ \frac{1}{2}O_2$	80 °C
$MgCl_2 + 2H_2O \rightarrow Mg(OH)_2$ $+ 2HCl$	350 °C

Table 2.5 presents, in more detail, some of the more important operating parameters of a few selected cycles. Note that the iron–chloride–oxide process refers to the Mark 9 cycle. A heat to work efficiency of 30% is assumed. Incidentally, this figure is important when comparison of thermochemical efficiency and electrolytic efficiency is made, as it sets the limit for hydrogen production via the latter route.

Hybrid thermochemical–electrochemical cycles

A very promising thermochemical cycle is that proposed early in 1978 by Westinghouse Corporation and which is known as the sulphur–iodine cycle. This is a hybrid cycle which employs electrolysis and high-temperature chemistry to decompose water. This company is presently evaluating the electrolysis of sulphurous acid at high temperature and pressure, developing materials for handling corrosive substances, and estimating the overall economics of the process. The Westinghouse cycle is not being developed solely for incorporation into a nuclear power plant, and a number of heat sources, including coal, are being considered.

The most critical problems of the process, figure 2.6, are involved with containing sulphuric acid at pressures of 300 psi and temperatures of up to 450 °C during the vaporisation steps before oxygen generation via sulphur

Hydrogen and Energy

Table 2.5 Process conditions for some water decomposition processes

	Caesium oxide process	Vanadium chloride process	Calcium bromide process	Fe–chloride– oxide process	HCl electrolytic process
Process heat/ kcal (mol H_2)$^{-1}$	125.0	155.0	88.3	66.0	13.8
Pumping/separation work/ kWh (lb H_2)$^{-1}$	1.32	0.90	18.2	not available	3.6
Waste heat/ kcal (mol H_2)$^{-1}$	12.5	161.0	4.5	63.5	121.5
Electrical work input/ kWh (lb H_2)$^{-1}$	—	—	—	—	15.3
Total energy input/ kcal (mol H_2)$^{-1}$	142.5	385.0	114.0	129.5	206.8
Thermal efficiency HHV	48%	18%	59%	53%	33%
LHV	41%	15%	49%	45%	28%
Highest endothermic reactor temperature/°C	1050	725	730	650	816
Fraction of process heat at highest temperature	70%	30%	26%	32%	7%
Reactions in closed cycle	4	4	4	5	2
H_2 delivery pressure/atm	(1)	(1)	15	1	19

Heat to work efficiency 30% whenever applicable

dioxide reduction. The capacity of such a plant is seen to be 380×10^6 standard cubic feet per day, at an overall efficiency of about 50%.

Primary energy sources for thermochemical processes
Thermochemical cycles were first conceived with the object of producing hydrogen from the thermal output of high-temperature gas reactors (HTGR), and obviously the technical development of such systems is of paramount importance as far as the potential of thermochemical splitting is concerned, since temperatures available from light water reactors have not the same capability.

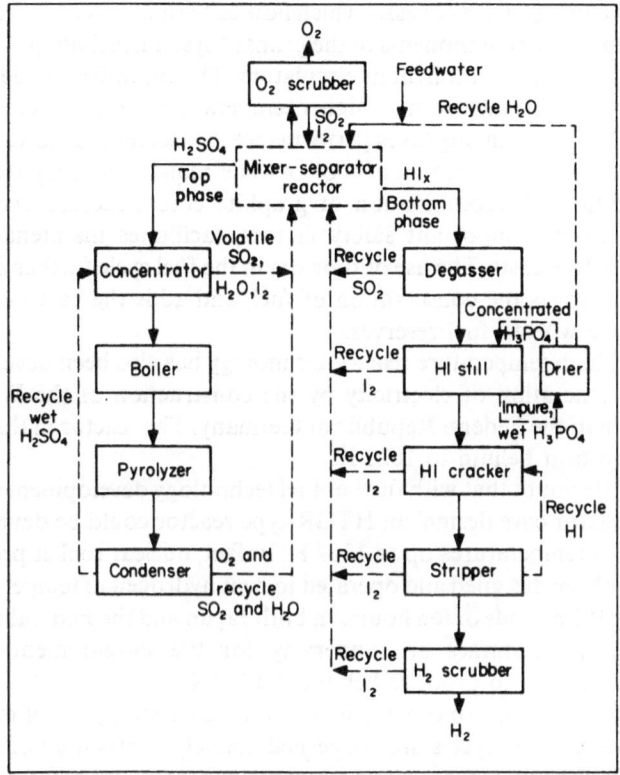

Figure 2.6 The process for the sulphur–iodine water splitting system
involving the reactions

$$2H_2O + SO_2 + I_2 \rightarrow H_2SO_4 + 2HI_x \quad \text{(aqueous 298 K)}$$
$$2HI_x \rightarrow xI_2 + H_2 \quad \text{(573 K)}$$
$$H_2SO_4 \rightarrow H_2O + SO_2 + \tfrac{1}{2}O_2 \quad \text{(1144 K)}$$

Adapted from *Chem. and Eng. News*, Nov. 21st, 27 (1977)

Although originally developed for the production of cheap electricity, the HTGR seems well suited to act as a heat source for the thermochemical production of hydrogen. Six large commercial plants, developed by General Electric, have been sold in the USA, and two more are on option. This strong penetration of the nuclear power market is the direct result of the successful operation of the prototype HTGR in Pennsylvania and the successful construction of a demonstration HTGR in Colorado.

The design of the commercial HTGR power plant is based upon the use of high pressure helium to transfer the heat generated in a graphite moderated uranium thorium core to an array of once-through steam generators located in the walls of a concrete reactor vessel which houses the entire system. This vessel contains all the major components of the primary system, including the reactor case, steam generators, and helium circulators. The incentive for developing this type of reactor results from its higher temperature[19] and therefore higher thermal efficiency, which improves performance, conserves fuel, lowers capital cost, and also reduces the amount of cooling water required to carry away waste heat. In addition, the combination of graphite core structure and helium coolant contributes important safety factors, facilitates maintenance, and helps decrease fuel costs. The use of thorium in the fuel cycle further decreases fuel costs, improves the conservation of fuel, and adds the vast deposits of thorium to the world's fuel reserves.

In Europe, high temperature reactor technology has also been developed for commercial generation of electricity by the construction of the WestFalen Power station in the Federal Republic of Germany. The reactor in this facility is designed to heat helium to 1023 K.

There is little doubt that with further fuel technology development and with a different reactor core design, an HTGR-type reactor could be developed to heat helium to temperatures up to 1367 K. In fact, nuclear rocket propulsion reactors have been designed and operated to heat hydrogen to temperatures of up to 2000 K for periods of ten hours. In both Japan and the Federal Republic of Germany, programmes are underway for the development of high temperature reactors to produce helium at 1270 K.

Besides nuclear heat, two other primary sources of energy are of interest as far as thermochemical cycles are concerned: namely geothermal and solar.

Geothermal heat appears to be unattractive, at present, since only temperatures below 250 °C are available from known sites. In addition, geothermal sources are characterised by sparse geographic distribution and limited energy capability, which means that an optimum plant size could not be achieved. On the other hand, solar energy is not subject to these constraints and, moreover, thermochemical production of hydrogen offers the goal of effectively converting solar energy into a usable form. Photo-thermal energy conversion is still an infant technology,[20] and present study is still basically at the conceptual stage. However, a recent publication[20] has explored configurations that might best exploit and use solar heat and suggests some approaches.

Research and development into thermochemical processes

Of the thirty or so cycles so far published only seventeen chemical elements are represented besides hydrogen and oxygen. Using computers and specifying only iron and chlorine as alternative elements, no less than 301 cycles have been made available to the research team at Ispra,[21] whilst at General Atomic, California, a master list of 710 compounds, composed of 56 elements is being investigated under constraints that the stoichiometric reactions contain no more than a total of five compounds, no more than three products or reactants, and all reactions are thermodynamically allowable. However, since it has been estimated[22] that only about 50% of the cycles available are superior to water electrolysis on an energy efficiency basis, future cycles should also be screened with this in mind.

Perhaps the most serious problem to be overcome, before any practical process can be demonstrated, is that of material compatability. At elevated temperatures corrosion is even more severe, and research in this field will have to take the form of screening various materials. In fact, at Ispra, exploratory corrosion tests for alloys and refractories have already been made.[9] It must be emphasised, however, that this research is in a very early stage of development and, because of the tremendous practical benefit to mankind, much more money should be devoted to it.

Economics of thermochemical processes
No accurate economic assessment of any thermochemical process has yet been made, although tentative suggestions have been put forward.[9] This is mainly due to the fact that experimental work has not yet advanced to the stage where pilot plants can be constructed and run. However, the general consensus of opinion, amongst people active in this field, is that thermochemical processes possess high potential for producing low cost hydrogen if acceptable efficiencies can be demonstrated.

Photolytic hydrogen production
Photolysis is the process by which a compound molecule is decomposed using the energy of incident light. The photon absorbed raises an electron into an excited energy state, thereby making it available for pairing with an electron from a neighbouring atom or molecule in an electron-pair bond, and by this process new molecules are formed. The decomposition of a molecule requires the breaking of molecular bonds; in the photolysis of water, light provides the necessary energy for the bond breaking. The net photochemical reaction can be summarised:

$$H_2O(\ell) + h\nu \rightarrow \tfrac{1}{2}O_2 + H_2.$$

In this process light energy of 68.3 kcal is absorbed per mole of water decomposed. This energy may be considered to be stored in the reaction products for subsequent use. Because water is transparent to visible light it can only be photolysed with light by the use of photocatalysts; the function of such a catalyst is to absorb the incident light energy. High absorptivity together with broad spectral activity are the most important factors in selecting 'photosensitisers'. Including the photocatalyst (A) in the reaction mechanism results in the following half reactions:

$$H_2O + h\nu + 2A \rightarrow 2 \text{ (reduced A)} + \tfrac{1}{2}O_2 + 2H^+.$$

$$2(\text{reduced A}) + 2H^+ \rightarrow 2A + H_2.$$

Note that this is a catalytic process and thus photocatalyst A is not consumed. Three types of photocatalyst are available: salts, semiconductors, photosynthetic dyes, and with all the three types the reactions which occur are oxidation–reduction reactions. There is available an excellent review of the different mechanisms involved.[23]

As well as being a diffuse source, much of the incident radiation does not contain energy equivalent to that needed to photolyse water. Figure 2.7

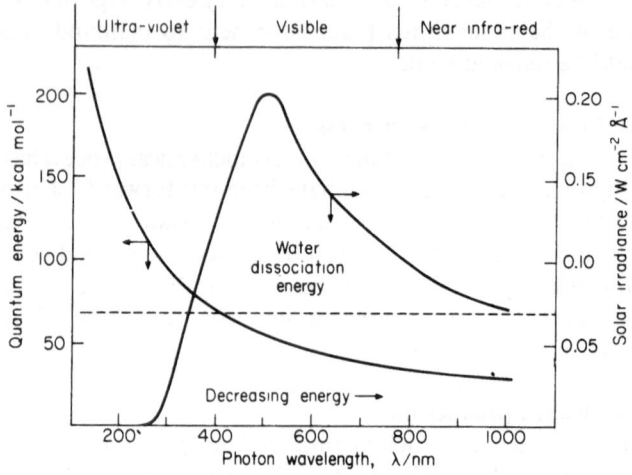

Figure 2.7 Energy content of incident radiation

compares the energy content of light quanta at different wavelengths with the energy needed to break the molecular bonds of water. It is clear from this figure that only ultra-violet light has the energy necessary.[24]

Photosynthetic organisms

It is possible to use photosynthetic organisms in a photochemical fuel cell. Plants and blue-green algae can use water as a reductant in light-dependent generation of compounds such as reduced ferredoxin and viologens. In living organisms, the reduced ferredoxin and adenosine triphosphate are then used to reduce carbon dioxide to cell material. But the production of cell material is not a necessary step in the harnessing of light energy and energy storage in the form of hydrogen would be more efficient and more direct.

In fact, certain plant forms are able to produce limited amounts of hydrogen, in addition to cell material. This phenomenon of hydrogen photoproduction was first observed in algae by Gaffron and Rubin in 1942[25] and in photosynthetic bacteria by Gest and Kamen in 1947.[26a] The blue-green alga of Benemann and Weare, *Anabaena cylindrica*, appears particularly promising, although much work needs to be done.[26b] Although early experimental work used freshwater species as subjects, the tropical and subtropical marine environments are better suited for studying such systems since they offer a greater range of species.

CHAPTER 3

TRANSMISSION AND DISTRIBUTION OF HYDROGEN

The movement of energy via pipeline is one of the cheapest methods of energy transmission, and since the system is buried underground, it is also acceptable environmentally.

As mentioned earlier, energy transport from its point of production to point of consumption will become a very important factor influencing any future energy system, and this is especially so in the case of nuclear energy. Generating plants of the future will be of far greater capacity than those in present-day use, and for safety reasons will probably be sited considerable distances from population centres.

Since electricity is the chief competitor of hydrogen as a means of carrying energy, transmission and distribution costs for both energy forms will have an important bearing in determining their relative competitiveness.

Natural gas transmission systems in most industrialised countries are well established and the technology highly developed. But, because of differences in physical properties, transmission of hydrogen in pipelines may slightly differ from that presently used for natural gas. This fact has an important bearing on whether or not existing natural gas lines can be modified to transport hydrogen. Since much experience has been gained during the US space programme in transporting hydrogen as a cryogenic liquid, the possibility of this method of transmission will also be examined.

Present status of hydrogen pipeline technology
The transmission of hydrogen under pressure and in large quantities is an established industrial practice but in most cases only very short distances are involved. In the main, hydrogen pipeline systems are confined to refineries where gas is moved from plant to plant and so far there has been little need or incentive to move hydrogen over great distances in this way.

Nevertheless, there are cases where hydrogen gas is transported over considerable distances by pipeline. Perhaps the best-known system is the pipeline network in the Ruhr area of Germany which is operated by Chemische Werke Huls AG and has a total length of 130 miles. The pipelines,

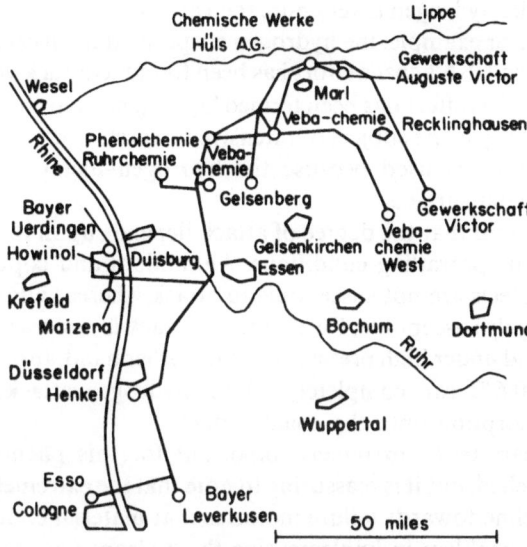

Figure 3.1 Map of the hydrogen pipeline of the Chemische
Werke Huls AG (1970)

having diameters ranging from 6 to 12 inches, are designed for pressures up to 240 psig. In 1972 the throughput was approximately 10.5 billion ft^3, or 3.4 $\times 10^{12}$ BTU. Figure 3.1 shows a map of the system.[27] Another hydrogen pipeline is located near Johannesburg, South Africa, and is approximately 50 miles in length.

None of the existing hydrogen pipeline systems are long enough to require compressor stations along their length. Hence, questions of optimum design parameters relating to compressor staging, cannot be answered by reference to the present systems. Theoretical models are being examined instead. However, experience in operating hydrogen pipelines can provide valuable information on other important aspects of transmission, such as safety and material compatibility.

Materials for hydrogen pipelines

Unlike methane, hydrogen has been known to interact with metals under certain conditions and this could well limit the materials suitable for construction of hydrogen pipelines. Existing mild steel pipelines have had a good record in this respect,[28] and only under conditions where atomic hydrogen is formed can penetration of the steel lattice occur. This phenomenon has been termed 'intergranular embrittlement'. Molecular hydrogen inside pipelines, at normal temperatures and pressures (< 2000 psi), will in most cases be inert.

However, if the hydrogen is very pure then attack at the surface of the metal will take place. For example, the hydrogen evaporated from cryogenic storage vessels used in the USA's space effort has been found to attack welded sections of these vessels. This effect has been termed 'hydrogen environment embrittlement' and presently is the subject of intensive research work by NASA.[29] The term 'environment' is used because the hydrogen–metal interaction only occurs at the metal surface.

It has been found that the degree of attack depends upon the nature of the metal and on the prevailing conditions. Aluminium and copper alloys and some stainless steels are not susceptible to attack whereas alloys of nickel or titanium are highly susceptible. Furthermore, attack is most severe at ambient temperatures and under high pressures. On the other hand, an oxygen impurity of as little as 0.6% can completely inhibit hydrogen attack, probably by preferential adsorption onto the metal surface.

It thus appears to be extremely important for this phenomenon to be carefully researched, but it is reassuring to note that the extremely good record of existing pipeline towards failure indicates that material compatibility may not be a major problem in implementing the hydrogen economy.

Modelling a hydrogen pipeline system

Design of a hydrogen pipeline system means specifying variables of the system such as diameter, thickness, and compressor power so that overall transmission costs are minimised. Analytical models are being constructed of such a system and should yield reliable results provided acceptable assumptions are made.

If existing unmodified gas lines are used to deliver hydrogen then these would be only capable of delivering some 29% of the original energy. This is mainly due to the lower volumetric heating values of hydrogen (about one third that of natural gas). Both compressor horsepower and capacity would have to be augmented in order that the energy capacity of the natural-gas line be restored to its original value. It is thus seen that evaluating hydrogen transmission costs under the above conditions would yield misleading results, since such a system is not optimised for hydrogen. Hence, it is essential to calculate from first principles the actual cost of moving hydrogen in a pipeline system fully optimised for hydrogen, and in a model in which the compressor stations, station spacings, and pipe diameter are all variable.

An analytical model for pipeline transmission of hydrogen has been developed by General Electric.[30] An example of how such a model is set up, of the cost assumptions made, and of the ensuing results will now be considered.

The purpose of a model is to minimise capital invested. Pipeline diameter is the variable used in the cost minimisation since it is the most important design parameter in determining operating characteristics and also because most of the capital cost terms can be expressed as diameter dependent quantities. The model is based on the equations for the isothermal flow of a compressible fluid

Table 3.1 Cost equations

$$C_{pipe} = 200\,W + 1300d + 70d + C_{row} \quad \text{(\$ per mile)} \qquad (1)$$

$$\text{where pipe weight } W = 28.2t\,(d-t) \quad \text{(tons per mile)} \qquad (2)$$

$$\text{pipe thickness } t = 0.7\,P_1 d/60\,000 \quad \text{(inches)}$$

$$C_{pump} = [230(\text{HP}) + 15\,000d]/L \qquad \text{(\$ per mile)} \qquad (3)$$

$$C_{trans} = 10^4 \left[\frac{0.15(C_{pipe} + C_{pump})}{(0.90)\,(8760)Q} + \frac{2546(\text{HP})}{0.40\,L\,Q}\,C_{fuel} \right] \qquad (4)$$

$$\text{(cents per } 10^6 \text{ BTU per 100 miles)}$$

and are a standard industry approach to preliminary pipeline design.[31]

Since the cost equations depend upon the various assumptions made, these are included in table 3.1. Equation (1) expresses the installed line cost as the sum of four terms: pipe costs at \$200 per ton times pipe weight per mile, installation at \$1300 per mile per inch of pipeline diameter, anti-corrosion wrapping and coating at \$70 per mile per inch of diameter, and right-of-way costs per mile. The first three terms depend on the pipe diameter, d, as computed by the pipeline model. Pipe weight per mile is given by equations (2) and (3): the former is a direct expression of pipe geometry, and the latter is the ASA code equation for the wall thickness of 60 000 psi minimum yield strength pipe in cross-counting application. Right-of-way costs are difficult to assess: a representative value of \$4000 per mile is used in the calculations. Since this quantity is independent of the pipeline parameters, its variation has no effect on the relative economics of the fuels considered. The cost data are based on average USA experience in relatively easy terrain for gas pipelines of 30 inches or larger diameter.[32] Figures are expressed in \$(1973) U.S.

Equation (4), which describes compressor costs, has a term proportional to installed power at \$230 per horsepower to represent the cost of land, buildings, contracts, etc., which are associated with each compressor station. The data are connected with pipe size at a value of \$15 000 per inch of outside diameter. Dividing single station cost by the interstation distance, L, expresses station cost on a per mile basis.

The sum of pipe and pump (compressor) costs in the investment are minimised by the model. Equation (5), which represents the operating cost, reflects a 15 % annual return on investment, a 90 % load factor for pipeline use, and the cost of fuel required to operate compressors. A nominal value of 40 cents per 10^6 BTU used for fuel costs in equation (5) is normalised by the energy flow rate, and gives the fuel transport costs in cents per 10^6 BTU per 100 miles.

Figure 3.2 illustrates the main results of the model. In a properly designed system, the cost of hydrogen transmission by pipeline is found to be between 30 and 50 % higher than for an equivalent amount of energy as natural gas.

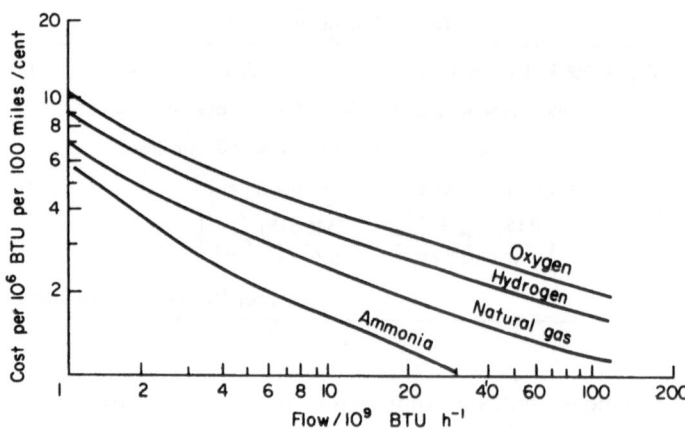

Figure 3.2 Pipeline transport costs of various fuels plotted against energy flow

Comparison of electrical and gas transmission costs

The quantity of energy a cable is able to carry is primarily limited by the amount of heat dissipated into the surroundings. Even in the case of overhead lines, where air serves as an effective heat sink, energy capacity is limited to about 2000 MW or 6.8×10^9 BTU per hour. On the other hand a 36 inch diameter hydrogen pipeline, under optimal conditions, could carry more than three times this amount.

One of the major drawbacks associated with future overhead transmission is the large areas of land needed to satisfy predicted electricity requirements. Thus, there has been much pressure put on the electric utility industries to consider underground transmission. However, the costs associated with underground transmission are very much greater than those for overhead transmission, mainly due to the costs of the cable which must be insulated and armoured against external damage. Also, cost disparity is compounded by the fact that cable capacity is considerably lower. These factors have led to the very high cost estimate for underground transmission of 177 cents per 10^6 BTU per 100 miles (figure 3.3).

Liquid-hydrogen transmission

As an alternative to transmission of hydrogen gas, its tranport in the form of a cryogenic liquid is also being considered. A good deal of experience has already been gained in handling this form of hydrogen in connection with the space effort in the USA. Truck trailers of 10 000 gallons and rail wagons of up to 34 000 gallons have been used to transport liquid hydrogen, but pipelining of liquid hydrogen has been limited to short runs at the production plants, and somewhat longer lines at launching areas. No technology presently exists for moving liquid hydrogen in bulk over large distances, and since transport costs were relatively unimportant during the space programme, existing lines are

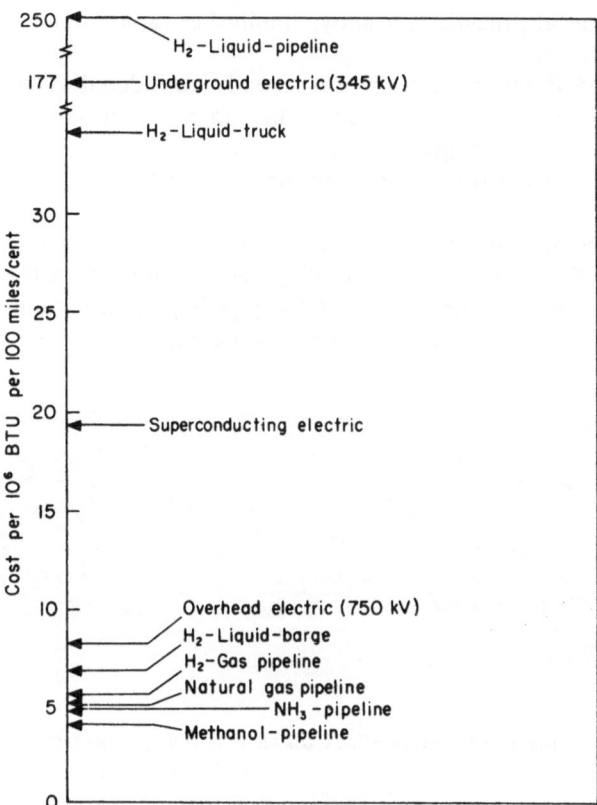

Figure 3.3 Relative energy transmission costs in the USA (1973), assuming 100 % use and 15 % fixed charge rate

very poorly insulated. The result is relatively poor performance and high costs (figure 3.2). In fact moving liquid hydrogen in a container by truck is considerably cheaper than moving it by pipeline, and transportation by barge is even cheaper.

Hence, it can be concluded that at present transport of hydrogen as a cryogenic liquid is economically unjustified, except in unusual circumstances; for example, where no liquefaction facilities are present for small-scale storage then hydrogen would have to be moved in this form.

Energy-pipe concept

Although too expensive to transport on its own it is possible that liquid hydrogen could be transmitted along with electricity in a common pipe. The very low temperature of the hydrogen (< 20 K) would be ideal for low-resistance electricity transmission. At such temperatures the electrical re-

sistance of certain metals and alloys diminishes rapidly and they become
super-conductors.

The electrical utility industry has already considered cryogenic transmission, but in most instances liquid helium has been considered as the
coolant. However, hydrogen also possesses suitable characteristics and in the
case of hydrogen more energy can be transported for a given capital
investment.

A proposed system is shown in figure 3.4. It consists of a steel pipe lined
internally with a super-conducting alloy and insulated both thermally and
electrically from the outer casing of the pipeline by an annulus of 'super-
insulation' (see reference 33 for more information).

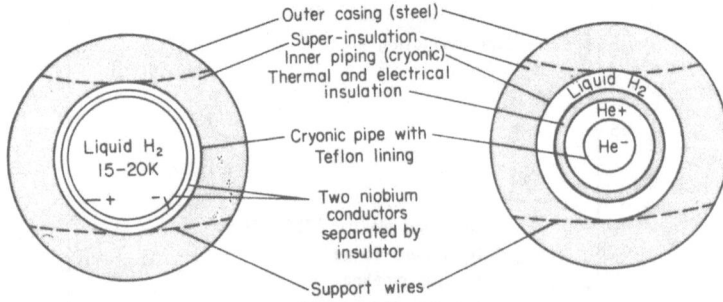

Figure 3.4 Two possible cross sections of an 'energy pipe'

CHAPTER 4

STORAGE OF HYDROGEN

If new sources of energy are to be fully exploited then an efficient energy storage system must be developed to meet variable demand. This is so because, in most cases, energy demands are periodic in nature whereas energy supply operates most efficiently on a constant output basis.

The supply and demand patterns of the electric utility industry illustrate the point well. Generating capacity must be sized for maximum demand, which can be more than twice minimum demand (see figure 4.1), since no economic storage method is currently available on a large-scale basis. The result is that fixed charges contribute significantly to electricity costs and consumers must pay heavily for a guaranteed supply.

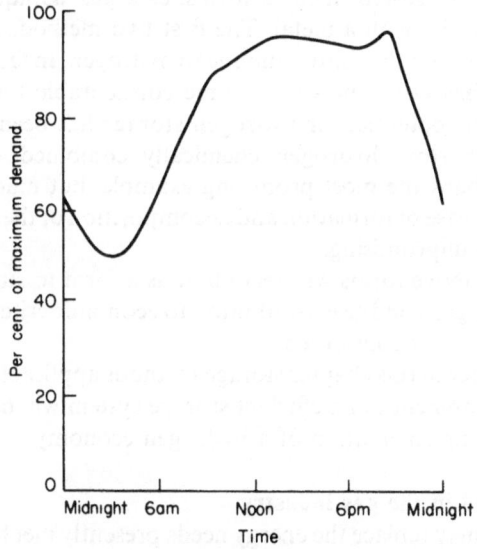

Figure 4.1 Typical daily load curve for electricity consumption

At present, the relative ease with which fossil-fuels are stored is taken for granted. The energy associated with such fuels is in the form of latent chemical energy which can only be released on combustion. On the other hand, in the case of new energy sources the actual form of heat is different. It is usually kinetic (wind, tidal) or heat* (nuclear, geothermal, solar) energy. Also it is not available uniformly, but rather on a cyclic basis, e.g. solar energy.

In order to enable effective storage, these new energy sources need to be converted into a secondary energy form. Electricity and hydrogen are the two most promising candidates to fulfil this role. However, electricity suffers from the disadvantage that it is almost impossible to store efficiently. To date, only hydro-electric storage systems have been utilised. Although these can be operated on a relatively high efficiency basis, their effective contribution to electricity storage is severely limited by the nature of the terrain required. Artificial reservoirs are far too expensive to be considered. Storage of electricity by means of batteries is not practical.

In contrast to electricity, hydrogen closely resembles our present fuels, especially natural gas. It can be made fluid and hence can be moved and stored in the same manner as today's fuels. Moreover, its energy is chemical in nature being released during oxidation of hydrogen to water.

As far as storing hydrogen is concerned, many useful comparisons can be drawn with the storage of natural gas, which is extensively practised in most industrialised countries, where as much as 75 % of demand in wintertime must be met from these stored reserves.

Hydrogen can be stored in three forms: as a gas, as liquid, or as a solid combined chemically with a metal. The first two methods are applicable to natural gas storage but the third is unique to hydrogen. In fact, the existence of metal hydrides has been known for some considerable time, but it is only recently that their potential for hydrogen storage has been recognised. It is even possible to store hydrogen chemically combined with non-metals; ammonia is perhaps the most promising example, but much would depend upon the relative ease of formation and decomposition of the chemical, and the economics seem unpromising.

Which of the above forms will serve best as a form to store hydrogen will depend upon the gas's end-use. In addition to economic criteria, safety aspects must also be carefully considered.

In the remainder of this chapter storage methods applicable to hydrogen are discussed. Development of an efficient storage system will have an important bearing on the implementation of a hydrogen economy.

Variable demand in the gas industry
Since hydrogen may replace the energy needs presently met by natural gas it is instructive to consider the present demand pattern. First of all, there very frequently exists an hourly variation in demand, which depends on such factors as extent of industrial load and time of day. In most instances the gas line capacity is such that variation can be met by alteration of line pressure, i.e.

---

* Heat energy is also kinetic energy, since it represents kinetic energy of individual atoms or molecules confined within a container.

linepacking. Secondly, in most distribution systems there occur brief periods of yearly maximum demand. Instead of sizing the main source of supply to accommodate these periods, it is far more economical to introduce supplementary supplies located close to load centres, e.g. liquified natural gas storage. Finally seasonal variations occur, dependent on the prevailing weather conditions. Depending upon the severity of winter, a seasonal variation factor of as much as seven can be experienced. Large scale storage is necessary to even out such variation, and reserves stored underground in depleted gas fields or aquifers are called upon.

Gaseous hydrogen storage

In comparison to underground storage, storage of hydrogen above ground on a large scale is expensive. The gas has an extremely low density (about 0.0052 lb/ft^3) and even at high pressures very large volumes will be required, resulting in high material costs. Alternatively if the storage space already exists, as it does in depleted gas or oil fields, use of such space would lead to low capital costs. Natural gas is stored in this manner, and in the USA storage of natural gas underground has reached the point where it is by far the largest storage capability in use. At the end of 1972, 337 gas storage locations were operational with a total capacity of some 5661 billion cubic feet, representing about 25% of the annual consumption of gas.[34]

The ability to store gas underground depends critically on the nature of the rock strata. Porous, permeable rock is required to hold, whilst sealing of the system is accomplished by the capillary action of water in the caprock. Certain areas are more suited for this type of storage than others and figure 4.2 shows which regions in the USA are geologically favourable for gas storage.

The pressure of gas existing within such an underground reservoir is not constant, but varies with depth as shown, figure 4.3. In many instances, natural gas is stored at greater pressure than that originally present in the field. This is the technique of 'overpressuring' and can increase storage capacity severalfold. However, certain limits on pressure gradients must be observed and it is generally accepted within the gas industry that pressure gradients of one psi/ft set a limit for storage. With this figure even as an upper limit one can still appreciate that substantial increases in storage capacity exist over and above the quantity of natural gas originally present (figure 4.3).

In addition to the use of depleted gas fields, it has been found that natural gas can be stored in porous rock simply by displacing the water present in the pores and creating artificial storage volume. This is the aquifer storage approach, and even though native gas or oil was never present it permits underground storage in sedimentary structures that have suitable caprock containment.

The pressure of gas necessary to displace the water will depend on the porosity and permeability of the sedimentary rock structures. In general, movement of gas–water interface is slow and at the end of the filling season

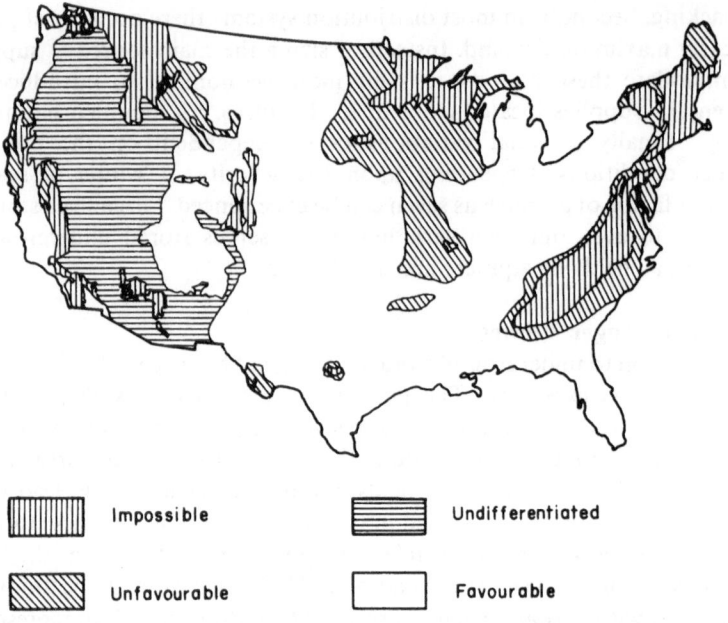

| Impossible | Undifferentiated |
| Unfavourable | Favourable |

Figure 4.2 Areas in the USA suitable for underground storage of gas in porous rock

(usually summertime), the gas pressure will exceed the formation water pressure and the interface will gradually recede. A minimum amount of gas must always remain in storage to prevent water from completely reoccupying the rock pores.

The volume of gas able to be stored via this technique will depend on the degree of rock porosity. In addition, acceptable rates of gas injection and withdrawal must be achieved. These are governed by the permeability of the rock. Three aquifer storage areas exist in France, the largest of which is at Beyners, near Paris, which has a capacity of seven million cubic feet.[35] Even in the absence of porous rock formations, storage volume can be created artificially, either by using mined cavities or by leaching out water-soluble minerals (e.g. salt) from the rock.

Since underground storage of natural gas has been operated successfully the question naturally arises of how compatible is hydrogen to such means of storage. Perhaps the most important difference between natural gas and hydrogen is the value of their respective diffusion coefficients. Under similar conditions, the hydrogen molecule is small enough to diffuse through an orifice three times more quickly than a natural gas molecule, and this can make sealing of hydrogen storage systems more difficult. But fortunately where gas is stored in the pores of rock, then sealing is achieved by capillary action of water, which fills all the voids in the caprock structure. Provided gas pressures

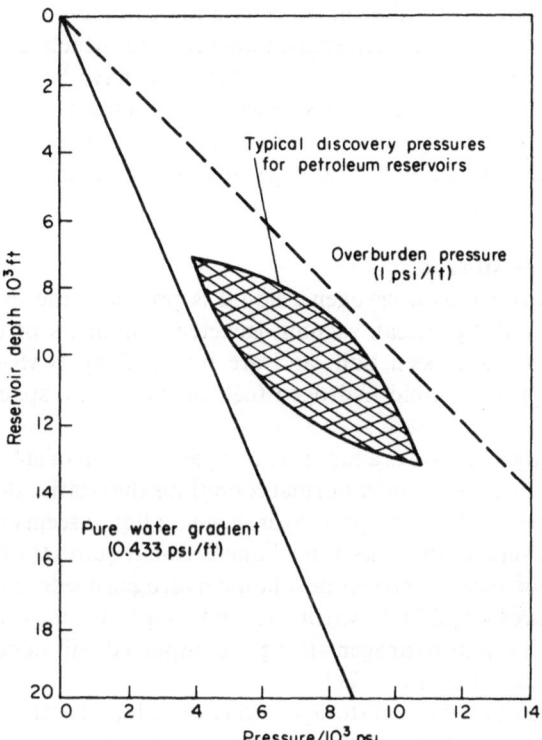

Figure 4.3 Depth against pressure relationships for underground reservoirs

are not exceedingly high, the caprock will act as an effective barrier to the
passage of any gas. Hence, in these cases sealing efficiency will be independent
of the gas diffusivity. Unlike depleted gas field and aquifer storage systems,
cavity/cavern storage involves large, open spaces to be filled with gas. Even in
the case of natural gas, effective sealing is a difficult obstacle to overcome, and
the problem would be greater for hydrogen. Nevertheless, the exploitation of
porous rock strata, presents no sealing problems and this is confirmed by the
successful storage of helium—a gas with similar diffusion characteristics to
hydrogen—underground in Texas.[36]

The storage potential of underground systems is enormous, with depleted
oil or gas reservoirs being economically most attractive for exploitation, since
man-made caverns involve excavation costs. In addition to this cost, there is
the requirement of effective sealing which can be difficult and costly.
Conventional mining costs are dependent on local geological conditions. In
easily sealed formations mining with nuclear explosives could lead to very low
capital costs but probably would be unacceptable for safety reasons.

Capital costs for underground storage are somewhat difficult to quantify because of the great variation in capacity and geological character of most systems. Much depends upon the maximum pressures which can be employed. Estimates ranging from $0.01 to $1.00 per kWh have been suggested. As mentioned earlier, above-ground storage in pressurised tanks is the alternative method for storing gaseous hydrogen on a large scale, but high capital costs make this system unfavourable; estimates as high as $2 to $3 per kWh have been put forward.[37]

Liquid hydrogen storage

Storage of hydrogen as a cryogenic liquid is presently the only large-scale method employed. Spherical, vacuum-jacketed containers of up to 900 000 gallons capacity have been built and are operated by NASA to ensure a continuous supply of liquid hydrogen fuel for the Apollo space programme (figure 4.4).

Liquid hydrogen provides a far more compact method of storage than does the gaseous form. In fact, under normal conditions the relative density factor is about 850. Superficially, storage of hydrogen as a liquid seems very attractive but there are several drawbacks. First, liquefaction requires the expenditure of large amounts of energy, and secondly liquid hydrogen needs to be kept at very low temperatures (< 20 K), which demands sophisticated insulation. The extent to which liquid hydrogen storage is employed will depend to a large degree on the end-use of the fuel.

Much background data on storage of cryogenic liquids can be derived from natural gas systems. Although practised on a somewhat smaller scale than gaseous storage, liquefied natural gas (LNG) storage is a rapidly growing technique. There are about eighty LNG storage facilities in the USA at the moment, and more are being constructed.

LNG is usually stored in flat-bottomed insulated tanks near load centres and is used to even out daily variations in demand, i.e. for peak-shaving purposes. It can also be employed to provide base-load supply, where facilities for pipeline transport do not exist. Many ships, designed to carry the cryogenic liquid, are currently in operation, and if the proposed scheme for importing natural gas into the USA goes ahead, shipping capacity will increase tremendously. Present use of LNG for peak-shaving purposes is likely to correspond to the role of liquid hydrogen in the hydrogen economy. However, base load usage of LNG involving as it does the transport of cryogenic liquid by ocean-going tankers will probably have no counterpart. Tankers will only be employed in areas where an off-shore pipeline is not feasible.

As hydrogen will be produced and transmitted as a gas then liquefaction facilities will be necessary at the place of storage. In fact, the process of liquefaction has very important bearings on the viability of liquid hydrogen storage.

Ideally, the necessary thermodynamic work required to take gaseous hydrogen at STP to the liquid state is approximately 1.8 kWh/lb with

Figure 4.4 A liquid hydrogen storage tank
(By kind permission of NASA, Washington, DC)

equilibrium conditions prevailing at each side of the process.[38] Figure 4.5 illustrates ideal refrigeration cycles for both natural gas and hydrogen. The above value of 1.8 kWh/lb represents about 10 % of hydrogen's heating value, but in fact about three times this amount of energy is actually required. This is because up to now the design of liquefaction plants has been primarily based on lowering capital cost and has not been concerned too much with efficient use of energy. Hence, there is room for improvement in this respect, and a target of 4.5 kWh/lb has been suggested as a possibility. Such an estimate would result in liquefaction energy requirements being 50 % higher than those of natural gas, on a pound for pound basis.

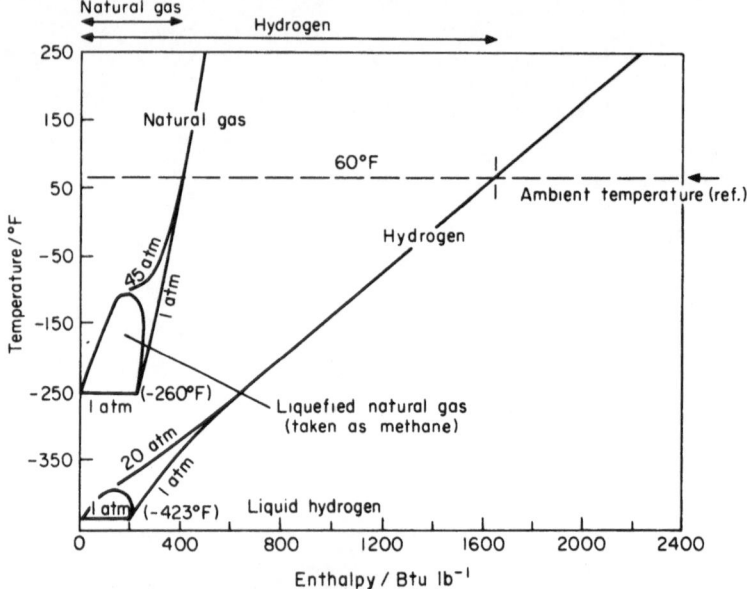

Figure 4.5 A comparison of the energy contents (per pound) of hydrogen and natural gas under conditions ranging from liquid to gases as ambient temperature

As far as capital costs of liquefaction facilities are concerned data available from NASA put such costs at $1.05 per 10^3 BTU/day.[39] This figure, on today's costs, would have to be increased, but not significantly if improved technology and economics of scale were fully exploited. However, it is the operating costs which control liquefaction economics, and hence efficiency improvement should be the major subject of development work. Depending upon the assumptions made, overall liquefaction costs can vary from $1 to $2 per 10^6 BTU.[40]

Metal hydride storage

It is well-known that certain metals and alloys can absorb hydrogen to form metal hydrides and furthermore that some hydrides contain far more hydrogen per unit volume than does liquid hydrogen.[41] However, it is only very recently that the concept of storing hydrogen in this form has arisen.[42]

The systems of interest react with hydrogen reversibly in as far as the gas can be recovered by lowering the pressure below, or raising the temperature above, the pressure and temperature of the absorption process. At a given temperature, each hydride is in equilibrium with a fixed pressure of hydrogen; its 'decomposition pressure'. If hydrogen is withdrawn and the pressure drops, then decomposition occurs until the evolved hydrogen has built up to the decomposition pressure again. This pressure is not only a function of the

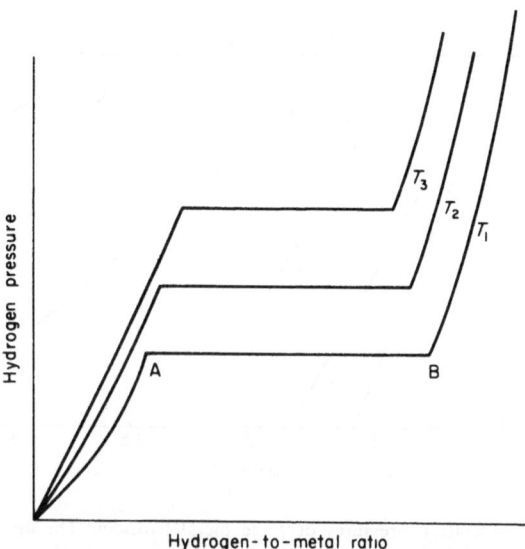

Figure 4.6 Pressure against composition isotherms for a typical hydrogen-metal system

temperature but also of the amount of hydrogen in the solid phase. This quantity is not usually constant but can often vary within rather wide limits. The way in which the dissociation pressure changes with the composition of the solid is shown in figure 4.6 for an idealised system M–H_2. As hydrogen is taken up by the metal and the ratio H:M increases, the equilibrium pressure increases rather steeply until a point A is reached. Up to this point, the solid consists of a solution of hydrogen in metal, rather than an actual compound. At higher concentrations, however, a second phase appears, having the composition B; and the addition of hydrogen does not result in an increase of pressure until all the solid phase has attained this composition. Above this 'plateau' region further enrichment of the solid in hydrogen requires a steep increase in pressure. The curves T_2 and T_3 illustrate the effect on the pressure against composition relation of raising the temperature. At temperatures above 300 °C, hysteresis is usually absent and the equilibrium pressure is the same, whether hydrogen has been added to or removed from the system. It is convenient to characterise the pressure against temperature relationship of a hydride system by reference to the well-known thermodynamic formula:

$$\log p = -A/T + B$$

where P is the pressure (atmospheres), T is the absolute temperature, and A and B are constants for any given system. Such an equation is valid over the normal range of pressures (figure 4.7).

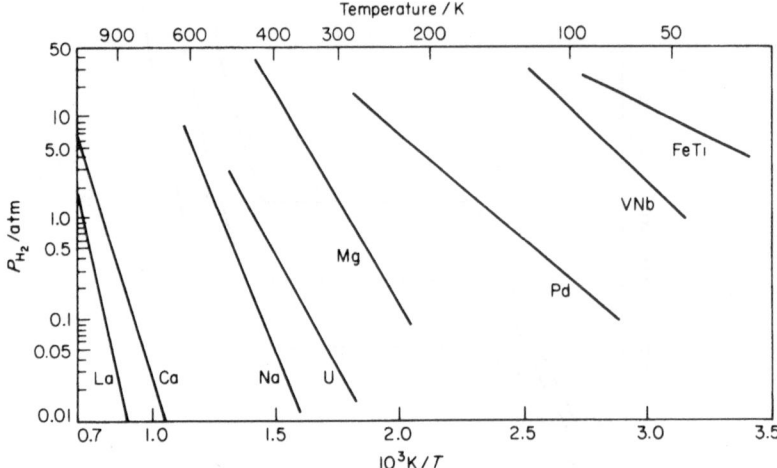

Figure 4.7 Dissociation pressures of some metal hydrides. The approximate compositions of the solid phases to which these curves refer are: LaH_2, CaH_2, NaH, UH_3, MgH_2, $PdH_{0.6}$, $VNbH_3$, FeTiH

What are the criteria which make a metal–hydride system attractive? First, and perhaps most important, the hydride must be thermodynamically stable with respect to its elements and also inert to reaction with either air or water. Depending upon the type of application, amount of hydrogen stored per unit weight and per unit volume are important characteristics. For transport high percentage hydrogen per unit weight and volume are essential. The quantity of heat needed to be transferred to decompose, or form the hydrides, will be governed by the enthalpy of decomposition. All those hydrides which are potentially useful are exothermic in nature, i.e. energy must be supplied to achieve their decomposition. Also of importance is the temperature at which the heat must be transferred to generate acceptable hydrogen pressure. For instance, hydrogen pressures of between 5 and 20 atmospheres are suitable for driving ground vehicles.

The rate of heat transfer is also important; the thermal conductivity of the hydride should be great enough to enable heat transfer to occur quickly. In the case of hydrides having high enthalpies of formation, correspondingly high values for thermal conductivity are required for the compound to be useful.

Besides thermal conductivity, the kinetics of hydride formation and dissociation will affect charging and discharging times, and it is the nature of the surface which influences rates of hydrogen absorption or desorption. High surface areas free from impurities are necessary.

Traditionally, research work on metallic hydrides has centred on equilibrium thermodynamic properties such as pressure–temperature–composition data and enthalpies of formation.[41] Much less data appear in the literature regarding reaction kinetics or thermal conductivity and this is probably due to the fact that techniques for studying such properties have only been introduced recently.

At several centres research is being conducted into the application of hydrides for fuel storage. No out-right winner has been found, although several promising hydride systems have been derived, and some are described briefly here.

Hydrides of magnesium and its alloys

Of the binary hydrides known before 1960, magnesium hydride was perhaps the most suitable for storage. It contains a relatively high percentage of hydrogen (7.65 %), decomposes at lower temperatures than most hydrides (one atmosphere of H_2 at 287 °C) and it is inexpensive. However, it suffers from the disadvantage that it does not form readily by direct combination of the elements: an over-pressure of many times the equilibrium dissociation pressure is required. Moreover, the enthalpy of decomposition is high.[43] Of perhaps most interest for the future is the compound derived from alloying magnesium with nickel, which enables the hydride $MgNiH_4$ to be formed.

Iron titanium hydride

Along with the hydrides of magnesium and its alloys, this particular system is the subject of intensive research at Brookhaven National Laboratories in the USA. The intermetallic compound FeTi reacts directly and reversibly with hydrogen to form iron titanium hydride according to the following equations:[44]

$$2.13\ FeTiH_{0.10} + H_2 \rightarrow 2.13\ FeTiH_{1.04} \qquad \Delta H = -6.7\ kcal$$
$$2.20\ FeTiH_{0.104} + H_2 \rightarrow 2.20\ FeTiH_{0.195} \qquad \Delta H = -7.5\ kcal$$

The products of the above reactions are grey metal-like solids which are very brittle but non-pyrophoric. The variation of pressure with hydrogen content is shown in figure 4.8. Note that high hydrogen pressures can be achieved at moderate temperatures, and there is another bonus because the heat of decomposition is less than half that of magnesium hydride, meaning that heat exchange will be achieved without difficulty. The chief disadvantage is that the elements involved have high atomic weights, and this hydride may only find use in storing hydrogen in a stationary facility.

AB_5 alloys

An interesting class of hydrides has been discovered by van Vucht and others at Eindhoven.[45] Certain alloys of the formula AB_5, where A is a rare earth metal and B is Fe, Co, Ni, or Cu can absorb up to seven hydrogen atoms per AB_5 unit. $LaNi_5$ absorbs six hydrogen atoms per formula unit at room temperature and at an equilibrium pressure of 1.5 atmospheres, forming the hydride

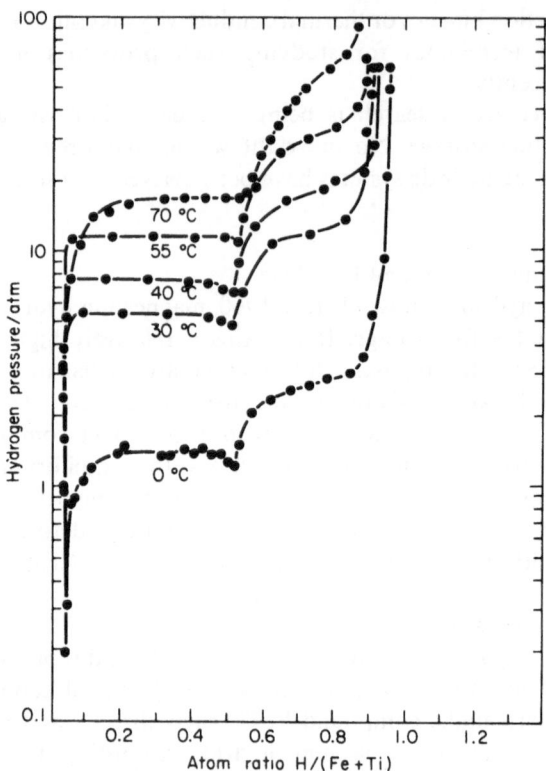

Figure 4.8 Pressure against composition isotherms for the
FeTi-H system

LaNi$_5$H$_6$. When saturated the hydride stores 1.3% of its own weight of hydrogen. LaNi$_5$ hydride exhibits several interesting properties. First, the sorption properties are relatively insensitive to impurities in the hydrogen. Secondly, the equilibrium pressure in the plateau region is nearly constant over the whole range and only slightly above one atmosphere at room temperature (figure 4.9). Finally, LaNi$_5$ alloy is relatively easy to activate.

Perhaps, one of the most important advantages of hydride storage is that the system is comparatively safe. The amounts of hydrogen released can be carefully controlled by varying temperature and pressure, and safety considerations will play a major factor in the use of hydrogen as a mobile fuel.

Other systems

At the University of Manchester Institute of Science and Technology McAuliffe has developed a series of simple manganese compounds of empirical formula MnLX$_2$ (L = tertiary phosphine, X = anion), which absorb/desorb molecular hydrogen at ambient temperature and pressure. These compounds appear to have great potential for facile hydrogen storage.

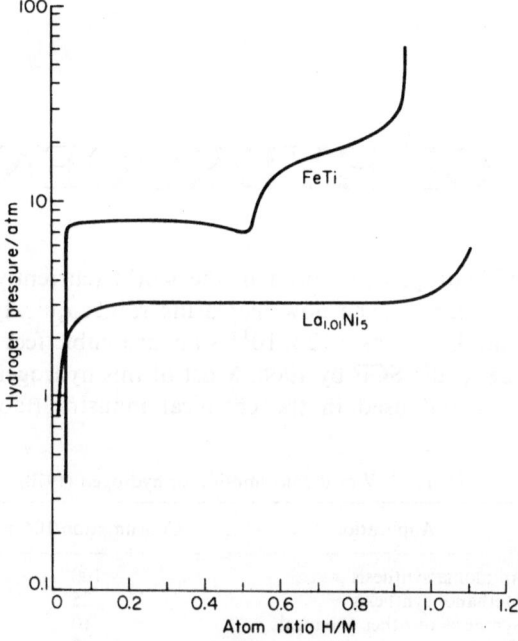

Figure 4.9 A comparison of the FeTi–H and La$_{1.01}$Ni$_5$–H systems
at 25 °C

CHAPTER 5

USES OF HYDROGEN

The amount of hydrogen produced in the world (currently more than 20 million tons per year) is even now increasing rather quickly. For instance, production in the USA was 7.22×10^{11} standard cubic feet (SCF) and had increased to 2.28×10^{12} SCF by 1968. Most of this hydrogen was produced from natural gas and used in the chemical industry (table 5.1) for the

Table 5.1 World consumption of hydrogen (1970)

Application	Consumption/10^9 m^3
Ammonia synthesis	100
Methanol synthesis	25
Synthesis of other chemicals	10
Hydrotreating/desulphurisation	30
Hydrocracking	30
Refinery fuel (low-grade H$_2$)	10
Total	205

production of ammonia, plastics, foodstuffs, rubbers and pharmaceuticals, and also as a reducing agent in the metallurgical and scrap-metal recovery industries. Occasionally there arises a special use for hydrogen, as in the recent use of liquid hydrogen in booster rockets for space vehicle launching.

I do not see a rapid revolution occurring, whereby hydrogen will quickly displace other primary and synthetic fuels. On the contrary, even before there is a 'hydrogen era' there will be a 'coal era', the latter lasting well into the next century. What is absolutely clear, however, is that the rate of growth in hydrogen production will increase, not only for the increased demands which exist as per table 5.1, but because the upgrading of the next, and final, batch of fossil fuels will need massive amounts of hydrogen. Table 5.2 shows some typical industrial hydrogen requirements, and it is obvious that these vital activities will lead to a rapid increase in demand for hydrogen. Whilst this occurs new methods, outlined in chapter 2, for hydrogen generation will be developed and, with increase in demand, more efficient and thus cheaper production will ensue. Various scenarios can be envisaged but, as well as being incorporated into fuels in upgrading and conversion of coal to liquid and

Table 5.2 Typical industrial hydrogen requirements

Use	Hydrogen requirement per unit of product/SCF
Ammonia synthesis	70 000–80 000/(ton of NH$_3$)
Methanol synthesis	36/(lb methanol)
Petroleum refining	$\geqslant$ 610/(bl crude oil)
Hydrotreating:	
naptha	50/bl
coking distillates	750/bl
Hydrocracking	2 000–2 500/bl
Coal conversion to:	
liquid fuel	6 000–7 000/(bl of synthetic oil)
gaseous fuel	~ 1 560/(10^3 SCF of synthetic gas)
Oil shale conversion to:	
liquid fuel	1 300/(bl of synthetic oil)
gaseous fuel	1 200/(10^3 SCF of synthetic gas)
Iron ore production	20 000/(ton of iron)
Process heat	3 070/10^6 BTU or
	2 700/(10^3 lb of process steam)

Table 5.3 Relative prices ($ per 10^6 BTU) for delivered energy (1970)
(US Federal Power Commission)

	Electricity	Natural gas	Electrolytic H$_2$
Production	2.67	0.17	3.00
Transmission	0.61	0.20	0.52*
Distribution	1.61	0.27	0.34*
Total	4.89	0.64	3.86

* Assumptions made.

gaseous hydrocarbons, hydrogen will also probably be reintroduced into the general gas supply by merely mixing it with natural gas. As for competition with electricity, table 5.3 shows how, even when produced by electrolysis, hydrogen is already in many respects a cheaper synthetic fuel than electricity.

In this chapter are outlined some of the more obvious uses of hydrogen, but before embarking on details of these it is important to consider the combustion properties of hydrogen and to establish how they compare with those of existing fuels (e.g. gasoline and Jet-A) and other synthetic fuels (e.g. methanol and ammonia).

It can be seen from table 5.4 that hydrogen has advantages and disadvantages when compared with natural gas (in table 5.4 it is assumed that CH$_4$ represents natural gas). For instance it is preferable to natural gas in the liquid form but suffers by comparison as a gas. Table 5.5 contains a more comprehensive comparison of fuels and it is clear that hydrogen is capable of carrying more combustible energy per unit mass than any other fuel, but

Table 5.4 Comparison of some physical and thermochemical properties of methane and hydrogen (data derived essentially from reference 6)

	Methane	Hydrogen
Gas density at 21° C and 1 atm		
lb ft^{-3}	0.0416	0.0052
g cm^{-3}	6.66×10^{-4}	8.33×10^{-5}
Liquid density at the normal boiling point		
lb ft^{-3}	26.53	4.43
g cm^{-3}	0.425	0.071
Liquid heating value/BTU lb^{-1}		
HHV	23 875	61 095
LHV	21 495	51 623
Gas heating value/BTU/SCF		
HHV	1012	325
LHV	911	275
Compressibility factor at		
1 atm	1.00	1.00
500 psia	0.935	1.020
1000 psia	0.873	1.065

delivers less energy per unit volume. But not only is the amount of energy released important when a fuel is burned, but there are other vital considerations. An obvious criterion to consider is the air/fuel ratio by weight and the volume per cent in air–fuel mixtures which will support combustion. Some data are collected in table 5.6 from which it is seen that hydrogen has remarkably wide combustion limits. This clearly makes combustion easier, and especially means that the necessity for tuning engines finely is removed; however, it also necessitates extra care against leakage or spillage because combustion is so much easier. Temperature is also an important factor in flammability limit considerations and these limits become wider at higher temperatures; flammability also depends upon ease of ignition (very low for hydrogen in comparison with methane). Not only is hydrogen easier to ignite, it is also much more difficult to quench once combustion is under way (as measured by the distance between two plates at which combustion ceases). The laminar burning velocity is a measure of flame propagation rates and this too is very high for hydrogen; the flame speed for hydrogen at ambient temperature and pressure is about 300 cm s^{-1}, whereas for methane + air, gasoline + air the corresponding flame speeds are about 35–45 cm s^{-1}. The main advantage of this exceptionally high burning velocity for hydrogen is that it allows for very small combustion chambers.

Schoeppel and his colleagues at the University of Oklahoma have made a thorough study of the internal combustion engine fuelled by hydrogen and have shown (figure 5.1) that the hydrogen + air mixture combusts to produce many

Table 5.5 Comparison data for some selected fuels J. A. Hoess and R. C. Stahman, *Unconventional Thermal, Mechanical, and Nuclear Low-Pollution-Potential Power Sources for Urban Vehicles*, Society of Automotive Engineers Transactions, Paper 690231 (1969)

Fuel	Density		Approximate energy per unit mass		Approximate energy per unit volume		Storage-system requirements for carrying the energy equivalent of 20 gallons of gasoline	
	g cm^{-3}	lb ft^{-3}	cal g^{-1}	BTU lb^{-1}	cal cm^{-3}	BTU ft^{-3}	weight/lb	volume/ft^3
Gasoline	0.735	45.9	10 700	19 100	7 840	876 000	138	2.9
No. 2 diesel fuel	0.838	52.4	10 600	19 000	8 900	995 000	138	2.6
Kerosene	0.821	51.3	10 700	19 100	8 670	970 000	129	2.6
JP-4 (jet fuel)	0.777	48.6	10 400	18 700	8 140	910 000	138	2.8
C$_2$H$_5$OH (ethanol)	0.789	49.1	6 500	11 600	5 130	570 000	221	4.4
CH$_3$OH (methanol)	0.795	49.7	4 900	8 700	3 860	432 000	294	5.9
CH$_4$(g)^b (methane)	0.136	8.5	12 000	21 500	1 640	183 000	515	28.5
CH$_4$(l)^c	0.425	26.5	12 000	21 500	5 100	570 000	248	16.6
NH$_3$(l)^a (ammonia)	0.635	39.7	4 500	8 000	2 840	317 000	469	13.8
C$_3$H$_8$(l)^a (propane)	0.579	36.2	11 100	20 000	6 470	724 000	202	5.7
C$_4$H$_{10}$(l)^a (butane)	0.563	35.2	11 000	19 700	6 220	695 000	193	5.9
H$_2$(g)^b (hydrogen)	0.017	1.1	29 000	52 000	500	56 000	1 370	92.0
H$_2$(l)^c	0.071	4.4	29 000	52 000	2 060	229 000	405	40.5
MgH$_2$^{d} (magnesium hydride)	0.870	54.4	2 070	3 710	1 800	202 000	731	12.0
VH$_2$^{e} (vanadium hydride)	6.400	400.0	610	1 100	3 920	440 000	2 300	6.0

^{a} Liquid at 27 °C. ^{b} Gas at 27 °C and 3000 psig. ^{c} Liquid at cryogenic temperatures and one atmosphere. ^{d} Magnesium hydride bed with 40% voids at 260 °C. ^{e} Vanadium hydride bed at 27 °C.

Table 5.6 Air-fuel stoichiometric ratios for support of combustion

Fuel	Stoichiometric air fuel ratio by weight[a]	Approximate volume percent of fuel in combustible mixtures near room temperature[b]
Gasoline	15.1	1.4 to 7.3
CH_3OH(methanol)	6.5	6.7 to 36.5
CH_4(methane)	17.3	5.0 to 15.0
NH_3(ammonia)	6.1	15.5 to 27.0
C_3H_8(propane)	15.7	2.1 to 9.4
C_4H_{10}(butane)	15.5	1.9 to 8.4
H_2 (hydrogen)	34.5	4.0 to 74.2

[a] Air is here assumed to be 21 % oxygen, 78 % nitrogen, and 1 % argon. [b] These data refer to flammability limits in air for upward propagation.

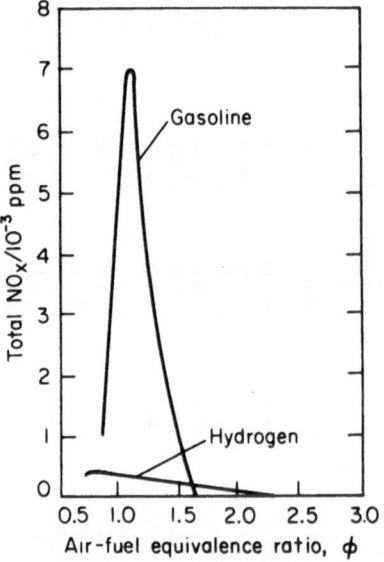

Figure 5.1 Wide-open throttle emission characteristics of single cylinder engines. [Reproduced, with modifications, from R. G. Murray, R. J. Schoeppel, and C. L. Gray, 'The Hydrogen Engine in Perspective', proceedings of the 7th Intersociety Energy Conversion and Engineering Conference, San Diego 1972 (published by the American Chemical Society)]

less NO_x pollutants than does the corresponding gasoline + air mixture. For poorly tuned engines it can be seen (figure 5.2) that in theory increased emissions of carbon monoxide and other organic compounds will result from a fuel-rich hydrocarbon + air mixture, whereas under such conditions the

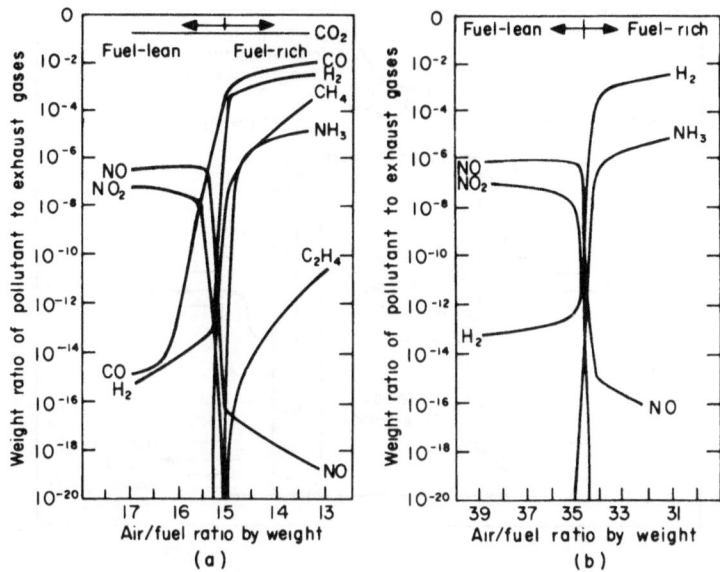

Figure 5.2 Theoretical pollutant emissions from (a) hydrocarbon- and (b) hydrogen-fuelled automobiles as functions of the air/fuel ratio by weight. [Reproduced from L. O. Williams, *Adv. Cryogenic Eng.*, **18** (1973)]

hydrogen + air mixture will yield decreased NO_x (but increased H_2 and NH_3) emissions.

It can thus be seen that hydrogen is generally superior to gasoline and natural gas in its combustion properties and, apart from its increasing use in the chemical and metallurgical industries (table 5.2), one can expect an increase in use for space-heating and cooking. Its use in transport will depend upon the development of cheap and efficient storage systems, except for use in aircraft where it has a very bright future (see later). Here I also wish to emphasise its use in fuel cells. Little more will be said here about combustion to generate shaft power except that there are essentially three ways in which this can be achieved with hydrogen: direct combustion with air in a conventional engine—either a reciprocating engine, a gas turbine, or a steam engine; by direct reaction with pure oxygen in a rocket engine—by injecting water, steam at almost any temperature and pressure can be generated in a completely enclosed no-intake no-exhaust system (except for water formed by combustion); and finally by direct electrochemical conversion in a fuel cell whereby electrical energy is generated for further conversion to shaft power in a conventional motor.

The hydrogen–oxygen fuel cell

In a hydrogen–oxygen fuel cell of the type illustrated in figure 5.3, hydrogen and oxygen gases enter at the anode and cathode, respectively, and at the anode

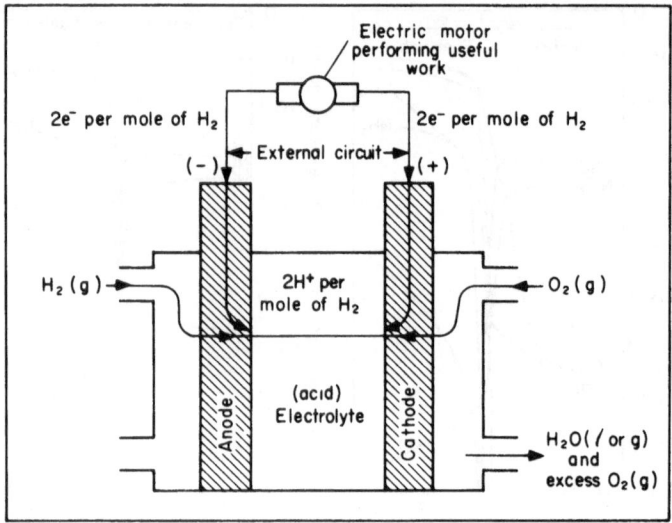

Figure 5.3 A hydrogen–oxygen fuel cell

the following reaction occurs:

$$H_2(g) \rightarrow 2H^+ + 2e^-$$

The protons migrate through the electrolyte to the cathode and the electrons are moved in the external circuit. At the cathode the following reaction occurs:

$$2H^+ + 2e^- + \tfrac{1}{2}O_2(g) \rightarrow H_2O(l)$$

The heat released in the overall reaction

$$H_2(g) + \tfrac{1}{2}O_2(g) \rightarrow H_2O \qquad (\text{at } 300 \text{ K})$$

is either 68.3 kcal or 57.8 kcal depending on whether the water formed is in the liquid or gaseous state.

The overall efficiency of a fuel cell can be expressed by the following

$$\eta_{FC} = \Delta\psi/\Delta H = \Delta F/\Delta H$$

where the useful work performed $(\Delta\psi)$ is equal to the Gibbs free energy change (ΔF) for the fuel cell reaction. Table 5.7 shows some data, which for all except the H_2–O_2 system are theoretical, for cells operating between 300 and 1000 K. The efficiency, η_{FC}, varies from 69 to 105%; efficiencies greater than 100% derive from the fact that heat can also be absorbed from the surroundings.

Fuel cells have much to recommend themselves for the future. In particular they do not exhibit the rapid decrease in efficiency with decreasing power output characteristic of most other types of power stations. Figure 5.4 illustrates this well for hydrocarbon–air fuel cells which have overall

Table 5.7 Fuel cell performances (Values for ΔH and ΔF are given per mole of fuel) [Reproduced with permission from *Thermodynamic Considerations of Fossil-Fuel Cells*, a paper presented by F. D. Rossini at the Seventh World Petroleum Congress, Mexico City (1967)]

Fuel-cell reaction	$-\Delta H$/kcal	$-\Delta F$/kcal	η_{FC}	ε/V
$H_2(g) + \frac{1}{2}O_2(g) \rightarrow H_2O(l)$:				
at 300 K	68.30	56.62	0.829	1.228
at 1000 K	- - - -	- - - - -	- - - - -	- - - - -
$H_2(g) + \frac{1}{2}O_2(g) \rightarrow H_2O(g)$:				
at 300 K	57.80	54.62	0.945	1.184
at 1000 K	59.21	46.03	0.777	0.998
$CO(g) + \frac{1}{2}O_2(g) \rightarrow CO_2(g)$:				
at 300 K	67.64	61.41	0.908	1.332
at 1000 K	67.55	46.67	0.691	1.012
$C(c, \text{graphite}) + O_2(g) \rightarrow CO_2(g)$:				
at 300 K	94.05	94.33	1.002	1.022
at 1000 K	94.32	94.61	1.003	1.026
$C_3H_8(g) + 5O_2(g) \rightarrow 3CO_2(g) + 4H_2O(g)$:				
at 300 K	488.5	495.7	1.015	1.075
at 1000 K	488.9	513.6	1.051	1.124

efficiencies of 40–50% for power output greater than 20 kW. For the hydrogen–air and hydrogen–oxygen systems overall efficiencies are 55% and 60%, respectively.

At the beginning of 1973 there were sixty 12.5 kW fuel cell stations under test in the USA under the auspices of the gas utility companies' TARGET (Team to Advance Research for Gas Energy Transformation) programme.

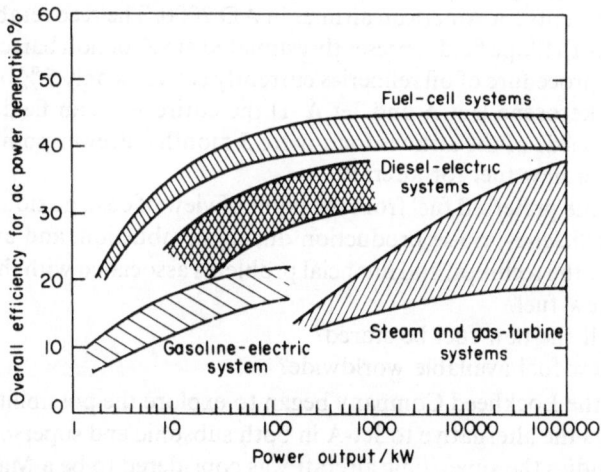

Figure 5.4 Fuel-cell efficiencies for hydrocarbon air systems producing $H_2O(g)$. [Reproduced from 'Hydrogen and other Synthetic Fuels', US Atomic Energy Commission, TID-26136, US Government Printing Office', Washington DC (1972) with permission]

Fuel cell stations working on hydrogen would be virtually non-polluting and could be sited even in dense urban areas. Transmission of gaseous hydrogen and oxygen to these stations would be inexpensive compared with transmitting electrical power to the same urban areas over considerable distances.

The potential for the hydrogen – oxygen fuel cell is enormous, and there is a further avenue along these lines which should be mentioned which illustrates the particular advantage which hydrogen offers as an excellent energy store/fuel. This is the attraction which hydrogen has as a solar energy storing device when such energy is captured from wind. Wind is a very attractive clean power source, but its capricious nature does require a system which is flexible and which allows for energy storage when the wind blows and the ability to release it more or less continuously when needed. A scheme has been outlined[46] in which wind energy is used to generate electricity to electrolyse water. The oxygen and hydrogen can be stored and fed at a constant rate into a fuel cell to produce electricity (figure 5.5).

Liquid hydrogen in aviation

It is becoming increasingly clear that an alternative to jet fuel (Jet A) will have to be found[46] and of all the uses for hydrogen none is as urgently required as that for fuelling aircraft. During the Arab oil embargo of 1973 the US government assigned commercial airlines a very low priority in allocation of fuel stocks, and the USA's airline industry has had a very hard look at the prospects of liquid hydrogen (LH_2) as a fuel and appears to like what it sees.[47 – 49]

Brewer[46] highlights the problem very well when he considers the projected fuel requirements for American airlines in AD 2000. The recoverable oil in the Alaskan North Slope field is presently estimated at 9.6 billion barrels. Since the processing procedure of oil refineries currently converts only 5 % of a barrel of crude into kerosene (Jet A and Jet A–1) the entire Alaskan field could only supply the needs of the airlines for a mere 3 months. Brewer points out some criteria for a new fuel for aircraft:

1. What is the preferred fuel from the point of view of cost, noxious emissions, energy efficiency, noise production during combustion, and availability?
2. What are the technical and financial problems associated with the transition to the new fuel?
3. How will the new fuel be stored?
4. Is the new fuel available worldwide?

In 1972 the Lockheed Company began to explore the possibilities of using hydrogen as the alternative to Jet-A in both subsonic and supersonic aircraft. In these studies the supersonic aircraft was considered to be a Mach 2.7 type, carrying 234 passengers and total payload of 49 000 lb and with a range of 4200 nautical miles; the subsonic passenger aircraft would carry 88 000 lb payload including 400 passengers over a range of 3000 or 5500 nautical miles. The aims of the study were to: (1) assess the feasibility of using LH_2 in commercial

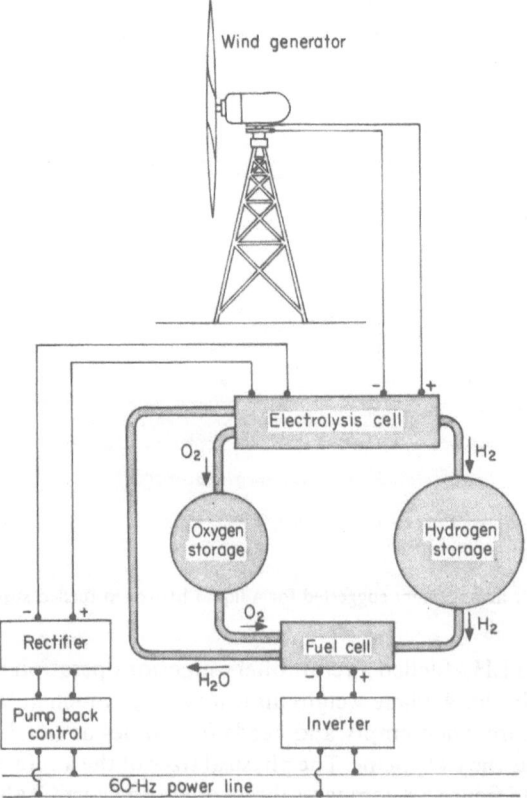

Figure 5.5 A suggested combined wind-electrolysis arrangement. Wind
provides the power for electrolysis of water, and the gases then produce
electric current in a fuel cell. Hydrogen and oxygen storage overcomes
the problem of variable wind supply

aircraft, (2) determine the advantages/disadvantages with respect to Jet A fuel,
(3) identify the technological problems associated with LH_2, and (4) outline a
plan for the development of this technology.

Operation by 1990 was the goal of the study and it was assumed that LH_2
was to be available at the airport at reasonable projected costs. Turbofan
engines based on advanced component technology were 'synthesised' by the
Lockheed computer for both LH_2 and Jet A fuels. Figure 5.6 shows an outline
of the supersonic transport and it can be seen that a unique feature of this
design is that there is no access from the passenger cabin to the flight deck.
Access between these was considered neither an advantage nor a disadvantage.

A summary of the LH_2-fuelled aircraft characteristics and those of the
equivalent Jet A-fuelled plane is contained in table 5.8, and the relative
advantages/disadvantages are contained in the column marked 'Factor'. It can

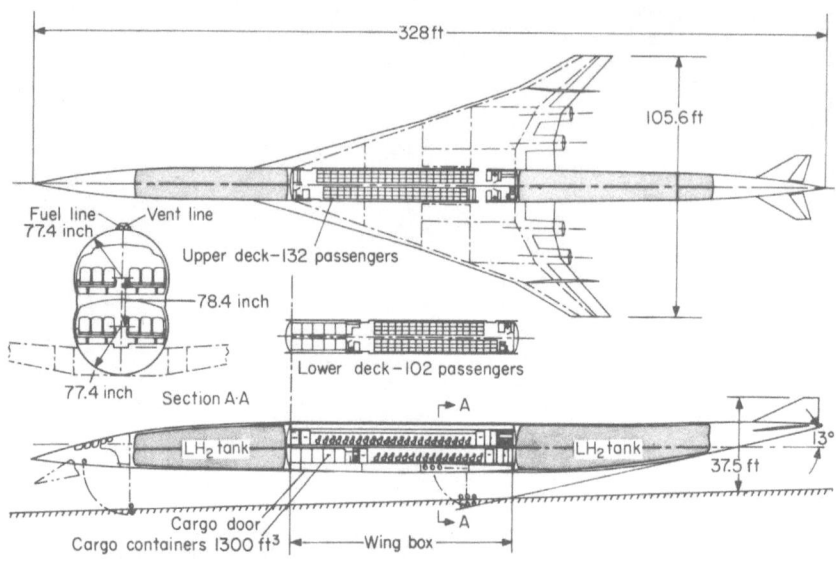

Figure 5.6 Interior arrangement suggested for a liquid hydrogen fuelled supersonic transport

be seen that the LH_2-fuelled aircraft offers superior operation in almost every category, e.g. the Jet A plane weighs almost twice as much at take off, weighs almost 40% more when empty and needs four times as much fuel to fly the same distance as the LH_2 plane. The physical sizes of the aircraft are compared in figure 5.7, and figure 5.8 compares direct operating costs. Table 5.9 contains some environmentally important parameters, from which it is again seen that LH_2 is a superior fuel to Jet A.

The comparison between LH_2 and Jet A for subsonic aircraft yielded similar results. The general conclusion for both types of aircraft is that LH_2 is much superior to Jet A as a fuel as it results in aircraft which are lighter, quieter, need smaller engines, can operate from smaller airports, cause less pollution and use up much less energy in their operation. The LH_2 aircraft would result in substantially reduced operating costs, table 5.10. These results suggest that LH_2 is indeed the air-fuel of the future.

Safety aspects of hydrogen use

No one could possibly deny that the production, transmission, storage, and use of hydrogen can be hazardous. Fuels, of their very nature, are hazardous. However, hydrogen seems to invoke more fears than almost any other portable fuel and this, for the most part, derives from the fire which destroyed the Hindenberg airship in 1937 at Lakehead, New Jersey. This fire, which has been communicated to two generations of people by eye-witness recordings and film of the disaster, has left a tremendous emotional effect which

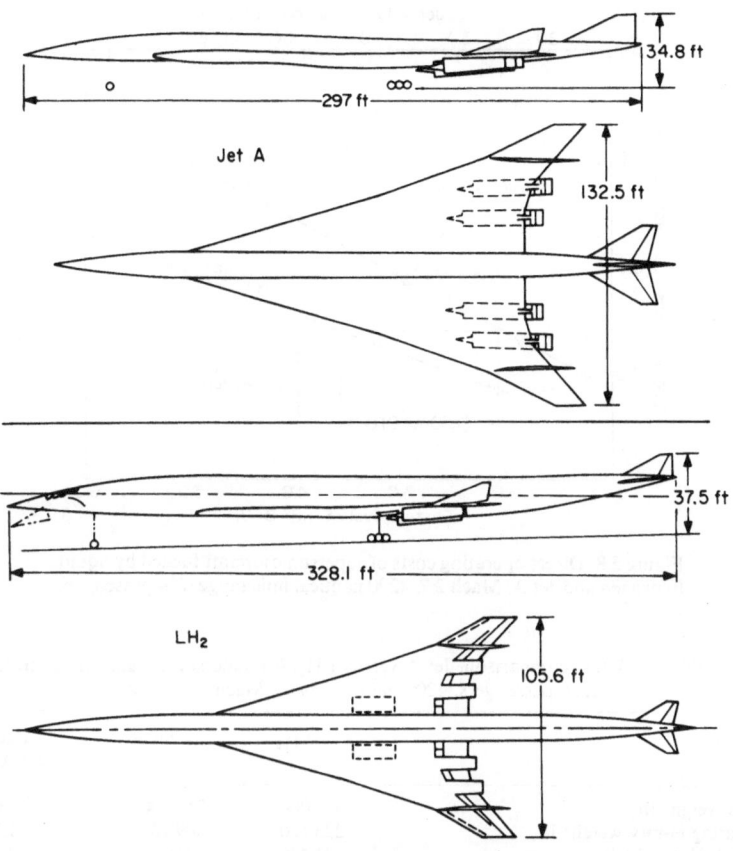

Figure 5.7 Size comparison between liquid hydrogen fuelled and conventionally fuelled supersonic transport aircraft

frequently makes for irrationality when the safety aspects of hydrogen are considered.

The Hindenberg fire should be put in a proper perspective. The ship was designed to run on helium and the gas-venting apparatus for bringing the ship down to land was placed inside the hull and the gas exhausted through air shafts out of the top of the ship. The ship had vented hydrogen for ten minutes during a thunderstorm in order to bring it near the ground. It was then moored to the ground by a steel mooring line (in an electrical storm!). Such a potentially dangerous situation would not be tolerated today.

However, one can point out one optimistic fact from this tragedy. The fire burned over seven million cubic feet of hydrogen in little more than one minute, killing thirty six people, *but sixty five others survived.* Hydrogen is lighter than air and burns *upwards*, whereas kerosene and Jet-A are heavier

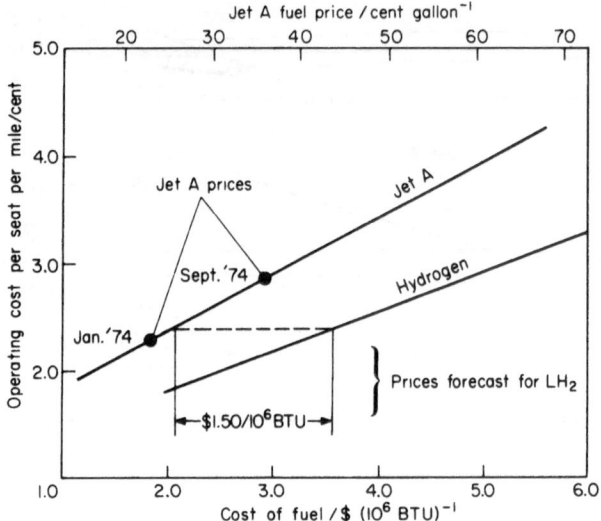

Figure 5.8 Direct operating costs of supersonic aircraft fuelled by liquid hydrogen and Jet A. Mach 2.7, 4200 nautical mile range, 234 passengers

Table 5.8 A fuel comparison, Jet A versus LH_2, for supersonic transport aircraft (234 passengers, 4200 nautical miles, Mach 2.7 speed)

	LH_2	Jet A	Factor (Jet A/LH_2)
Gross weight/lb	368 000	750 000	2.04
Operating empty weight/lb	223 100	309 700	1.39
Block fuel weight/lb	81 440	326 000	4.00
Thrust per engine/lb	46 000	89 500	1.94
Span/ft	105.6	132.5	1.25
Height/ft	37.5	34.8	0.93
Fuselage length/ft	328	297	0.91
Wing area/ft^2	6880	10 822	1.58
L/D (cruise)/(lb h^{-1})lb^{-1}	6.99	8.5	1.21
SFC (cruise)/(lb h^{-1})lb^{-1}	0.561	1.51	2.69
Aircraft price/10^6\$	48.0	67.3	1.40
Energy use/BTU seat^{-1} (nautical mile)$^{-1}$	4274	6102	1.43

than air and remain for long periods in the vicinity of burning aircraft.

Hydrogen explosions are rare. So-called 'town gas' has typically contained fifty per cent hydrogen without any report of undue difficulties. But of course all new forms of energy are met with stiff codes of handling and often irrational fears. Witness the law which required automobiles in England to be preceded by a man carrying a red flag. In a similar manner the first tank trucks of liquid

Table 5.9 Environmental acceptance parameters for supersonic transport aircraft. The data refer to a General Electric J85 engine in simulated flight at Mach 1.6 and at 55 000 feet

	LH$_2$	Jet A
Environmentally perceived noise/dB		
sideline	105.9	108.0
flyover	104.3	108.0
Sonic boom overpressure/16 ft^{-2}		
start of cruise	1.32	1.87
end of cruise	1.19	1.40
maximum encountered (during climbout)	2.08	2.50
Exhaust emissions		
NO$_x$/g kg^{-1}	Low	3.7*
CO/g kg^{-1}	None	90*
unburned hydro-carbons/g kg^{-1}	None	0.5*
odours	None	Objectionable
H$_2$O/lb (nautical mile)$^{-1}$	146	75.3

Table 5.10 Summary of characteristics for supersonic transport aircraft. A cross indicates which fuel is preferable in each case

Characteristic	LH$_2$	Jet A
Weight	X	
Noise	X	
Pollution	X	
Energy use	X	
Production price	X	X^a
Operating costs for fuel	X	
Engine size	X	
Physical size	X	X^a
Small airport capability	X	

^{a} Subsonic

hydrogen on American roads in 1956 were escorted at the front and rear by red jeeps. Contrast this with the present position: liquid hydrogen is shipped by road and rail fairly routinely, giving rise to no more concern than does the transport of other flammable materials.

The problems of safe-handling should not be minimised, however. Hydrogen is colourless, odourless, tasteless, and non-toxic under normal conditions. It has a low viscosity, is volatile, and forms potentially explosive and easily ignited mixtures with air. On the other hand, leakages of hydrogen disperse rapidly, and this is a favourable property if adequate ventilation is allowed for. Table 5.12 compares the properties of hydrogen and methane

Table 5.11 Stoichiometric air fuel ratio by weight and the volume per cent of fuel in air fuel mixtures

Fuel	Stoichiometric air fuel ratio by weight[a]	Approximate volume per cent of fuel in combustible mixtures near room temperature[b]
Gasoline	15.1	1.4 to 7.3
CH_3OH (methanol)	6.5	6.7 to 36.5
CH_4 (methane)	17.3	5.0 to 15.0
NH_3 (ammonia)	6.1	15.5 to 27.0
C_3H_8 (propane)	15.7	2.1 to 9.4
C_4H_{10} (butane)	15.5	1.9 to 8.4
H_2 (hydrogen)	34.5	4.0 to 74.2

[a] An air composition of 21 % oxygen, 78 % nitrogen, and 1 % argon is assumed.
[b] The data refer to flammability limits in air for upward propagation.

Figure 5.12 Physical and thermochemical data for CH_4 and H_2; based on data from standard physical and thermochemical compilations and from D. P. Gregory *et al.*, *A Hydrogen-Energy System*, American Gas Association, Arlington, Virginia, 1973

	CH_4	H_2
Gas density at 21 °C and 1 atm/		
lb ft^{-3}	0.0416	0.0052
g cm^{-3}	6.66×10^{-4}	8.33×10^{-5}
Liquid density at the normal boiling point/		
lb ft^{-3}	26.53	4.43
g cm^{-3}	0.425	0.071
Liquid heating value/BTU lb^{-1}		
HHV	23 875	61 095
LHV	21 495	51 623
Gas heating value/BTU SCF^{-1}		
HHV	1012	325
LHV	911	275
Compressibility factor at		
1 atm	1.00	1.00
500 psia	0.935	1.020
1000 psia	0.873	1.065

which are of interest when safety is considered. Hydrogen leaks are more likely to occur, but in a confined space only about one quarter of the energy is available for fire or explosion.

Liquid hydrogen can cause liquefaction of oxygen, which is very dangerous, and thus any pipes or tanks which would be used to carry LH_2 would have to be adequately purged with either hydrogen or helium prior to filling. Overall, however, it must be emphasised that few accidents have recently been reported. The US Bureau of Mines[50] has caused LH_2 spills in order to examine the rate of hydrogen cloud growth, cloud volume, and detonation potential. Such spills ranged from 1 to 5000 gallons (3.1 to 15.5×10^4 BTU). No detonations occurred during unconfined tests when ignition was attempted by an electrical discharge, although detonations did occur in confined spaces. Thus, provided adequate precautions are taken it seems that hydrogen, both as gas and liquid, may be used with as much safety as our present volatile fuels; ventilation seems particularly important in order to make use of the favourable ease of dispersal associated with hydrogen.

CHAPTER 6

ALTERNATIVE SYNTHETIC FUELS

We have seen that the production of hydrogen requires a net input of energy, but because of the physical and chemical properties of hydrogen, a total systems analysis may reveal that its use can lead to an overall conservation of energy when the total production–delivery–use 'cycle' is considered. Moreover, its use as a fuel can lead to a conservation of existing hydrocarbon resources and, because hydrogen fits well into current technological fuel manipulations, its use can lead to conservation of capital and economic resources.

However, hydrogen (or electricity) is not the only synthetic fuel which is attractive and needs to be considered seriously. Any alternative synthetic fuel must be capable of production from readily accessible raw materials, thus limiting consideration to those fuels constituted from elements present in the atmosphere, oceans, and biosphere. The fuels so far considered are ammonia, hydrazine, methanol, and ethanol, and of these methanol is by far the most attractive and rivals hydrogen as a multipurpose fuel for the next century. Indeed it is not unlikely that both hydrogen and methanol will find considerable use and that one will not exclude the other.

Air is a source of nitrogen used to make ammonia and hydrazine and could be a store for CO_2 to produce carbon fuels. Carbon dioxide is also available from limestone and seawater, but of course would be quite expensive to recover in large quantities. It is of significance that hydrogen is an essential precursor to any of the above mentioned fuels (excluding ethanol produced through an initial photosynthesis mechanism).

Table 6.1 gives a comparison of some of the important properties of these fuels with hydrogen and gasoline. While all these fuels have lower gravimetric heats of combustion than hydrogen they do offer some attractive physical properties, for example, higher volumetric heat of combustion and easier storability. They also generally have narrower concentration (in air) limits of combustion and detonation, and higher values of ignition energy. Some metals may be considered as synthetic fuels, since in their elemental form they can combine with oxygen and give off heat. The oxides can then be recycled for reduction back to their elemental state. Thus, as a metal, energy can effectively be stored and transported. Perhaps a more efficient system, however, is to consider metals as an electrochemical fuel, as in a rechargeable battery. Since

Table 6.1 Combustion energies per unit mass and per unit volume and approximate fuel and storage system weight and volume requirements for selected fuels; based on data from standard thermochemical compilations and from J. A. Hoess and R. C. Stahman, *Unconventional Thermal, Mechanical, and Nuclear Low-Pollution-Potential Power Sources for Urban Vehicles*, SAE Transactions, (1969)

Fuel	Density		Approximate energy per unit mass		Approximate energy per unit volume		Storage-system requirements for carrying the energy equivalent of 20 gallons of gasoline	
	$g\,cm^{-3}$	$lb\,ft^{-3}$	$cal\,g^{-1}$	$BTU\,lb^{-1}$	$cal\,cm^{-3}$	$BTU\,ft^{-3}$	weight/lb	volume/ft^3
Gasoline	0.735	45.9	10 700	19 100	7 840	876 000	138	2.9
No. 2 diesel fuel	0.838	52.4	10 600	19 000	8 900	995 000	138	2.6
Kerosene	0.821	51.3	10 700	19 100	8 670	970 000	129	2.6
JP-4 (jet fuel)	0.777	48.6	10 400	18 700	8 140	910 000	138	2.8
C_2H_5OH (ethanol)	0.789	49.1	6 500	11 600	5 130	570 000	221	4.4
CH_3OH (methanol)	0.795	49.7	4 900	8 700	3 860	432 000	294	5.9
$CH_4(g)^b$ (methane)	0.136	8.5	12 000	21 500	1 640	183 000	515	28.5
$CH_4(l)^c$	0.425	26.5	12 000	21 500	5 100	570 000	248	16.6
$NH_3(l)^c$ (ammonia)	0.635	39.7	4 500	8 000	2 840	317 000	469	13.8
$C_3H_8(l)^c$ (propane)	0.579	36.2	11 100	20 000	6 470	724 000	202	5.7
$C_4H_{10}(l)^c$ (butane)	0.563	35.2	11 000	19 700	6 220	695 000	193	5.9
$H_2(g)^b$ (hydrogen)	0.017	1.1	29 000	52 000	500	56 000	1370	92.0
$H_2(l)^c$	0.071	4.4	29 000	52 000	2 060	229 000	405	40.5
MgH_2^d (magnesium hydride)	0.870	54.4	2 070	3 710	1 800	202 000	731	12.0
VH_2^e (vanadium hydride)	6.400	400.0	610	1 100	3 920	440 000	2300	6.0

[a] Liquid at 27 °C. [b] Gas at 27 °C and 3000 psig. [c] Liquid at cryogenic temperatures and one atmosphere. [d] Magnesium hydride bed with 40% voids at 260 °C. [e] Vanadium hydride bed at 27 °C.

the combustion of metal fuels would divert supplies (of metal) from more important end-uses, their inclusion here is more of academic interest.

Each fuel will now be considered in greater detail. Table 6.2 contains an estimate of fuel costs in 1972.

Table 6.2 Production costs of energy contained in selected fuels. The costs apply to large plant capacities, assuming that 15 % of the plant cost is allocated annually to profit, interest, depreciation, and maintenance

Fuel	Source	Cost/$(10^6 \text{ BTU})^{-1}$
Gasoline	Crude oil	1.05
Methanol	Natural gas[a]	1.58
	Coal[b]	1.48
	Lignite[c]	1.25
Ethanol	Petroleum feed stocks	4.60
Ammonia	Natural gas[d]	1.57
Hydrazine		21.00
Methane	Well-head NG	0.15–0.40
	Imported LNG	0.80–1.00
	Coal	0.80–1.00
Hydrogen gas	Natural gas[a]	0.97
	Coal[b]	1.32
	Lignite[c]	0.78
Liquid hydrogen	Liquefaction	1.50[e]

[a] Natural gas at $0.40 per 10^3 SCF. [b] Coal at $7 per ton or $0.27 per 10^6 BTU. [c] Lignite at $2 per ton or $0.15 per 10^6 BTU. [d] Natural gas at $0.45 per 10^3 SCF. [e] Additional liquefaction cost.

Ammonia

Ammonia is produced today by direct catalysed synthesis from hydrogen and nitrogen (2000–5000 psi and 200–500 °C). The basic process has been in commercial use for many years and is considered to be a highly developed technology. Plant capacities have increased from 300 ton/day in the 1950s to around 1700 ton/day at present, with the effect that ammonia costs have virtually halved. Part of the cost decrease is also attributable to improved technology, including improved and larger compressors, improved catalysts, use of higher pressures, and vastly improved heat recovery systems. In addition, present capacity results in great competition. It is doubtful whether further cost reductions greater than 5 % to 10 % can be achieved. The present process is so cheap and efficient that it is also very unlikely that it will be supplanted by alternative methods now being evaluated, including biological production, use of transition metal complexes, thermochemical methods, or magneto-hydrodynamic generation; in fact, the latter two methods appear to produce only nitrogen oxides and not ammonia. World production capacity is around 60 million ton/year, with predicted annual growth rate in the region of 6 %.

Future costs of raw material hydrogen will be the main factor controlling prices of ammonia. It seems apparent that coal will be the agent for splitting water in the intermediate term whilst water electrolysis from highly efficient cells will compete effectively in the longer term.

At present, ammonia is mainly used as a chemical fertiliser in the form of urea and ammonium nitrate, phosphate, and sulphate. Production is mainly limited to those areas where sources of natural gas exist or where refineries are sited, and liquid NH_3 is piped to main agricultural areas. Since it can be readily stored and transported it can be considered as a portable synthetic fuel. It has been extensively evaluated as a fuel for internal combustion engines and for fuel cells but due to its toxic nature and relatively low heat of combustion, it has not found commercial application.

Properties of ammonia make it an attractive proposition as an energy storage medium but not as a fuel in its own right. It can be stored as a liquid under pressure at ~ 150 psi at ambient temperatures and its high density (relative to hydrogen) means that an equivalent amount of energy can be stored in about 60% of volume of liquid hydrogen. Equally important, ammonia can be readily transported in pipelines, barges, etc. more economically than can hydrogen. As mentioned above, the technology for moving and storing ammonia is already in existence in many parts of the world.

However, ammonia's relatively poor ignition and combustion properties mean that burning the compound as a fuel is not really feasible on the tonnage scale envisaged. In such a system atmospheric nitrogen would be slowly converted to toxic NO_x and NH_3. Whether ammonia can prove useful as a transport and storage system for energy can only be ascertained when a complete systems analysis is undertaken.

Hydrazine

Two processes are used for the commercial production of hydrazine: the Raschig process and the urea process. In the Raschig process the synthesis of hydrazine from ammonia and sodium hypochlorite takes place in two steps:

$$NH_3 + NaOCl \rightarrow NH_2Cl + NaOH$$

Sodium hypochlorite is manufactured by reacting Cl_2 with NaOH, both of which are products of brine electrolysis. Chloramine (NH_2Cl) is then reacted with excess ammonia to form hydrazine.

$$NH_2Cl + NH_3 + NaOH \rightarrow NH_2NH_2 + NaCl + H_2O$$

The first reaction proceeds rapidly, whereas the reaction of chloramine with ammonia is slow and rate determining, and requires heat. Chloramine may also react with hydrazine to form ammonium chloride and nitrogen.

$$NH_2NH_2 + 2NH_2Cl \rightarrow 2NH_4Cl + N_2$$

The reaction is carried out at elevated temperatures (130 °C) which favour hydrazine formation, effectively minimising the decomposition of hydrazine by chloramine, which is independent of temperature. It is necessary to use a large excess of ammonia (20:1 to 30:1). The hydrazine-containing solution is pumped to a crystallising evaporator, where the sodium chloride and sodium hydroxide are removed, and then to a fractionating column for water removal.

In the urea process, hydrazine is formed by the reaction of sodium hypochlorite and urea according to the following equation.

$$NH_2C(O)ONH_2 + NaOCl + 2NaOH \rightarrow NH_2NH_2 + NaCl + Na_2CO_3 + H_2O.$$

Two mixtures are prepared for feed to the reaction vessel. The first is prepared by chlorinating a 30% solution of sodium hydroxide until it has an available chlorine content of 140–155 g/litre and a residual sodium hydroxide content of 170–190 g/litre. The second solution, which is 43% urea, is prepared by dissolving urea in water while steam is passed through the solution to maintain the temperature at about 5 °C, since the dissolution is strongly endothermic. Four volumes of the first solution and one volume of the second to which has been added 500 mg of glue per litre are fed continually to the reactor, where the temperature is allowed to rise to 100 °C. The hydrazine is removed in the same manner as in the Raschig process.

Both the Raschig and urea processes are used at present for the production of hydrazine. The two processes are competitive because the more costly raw materials for the urea process are offset by the capital cost of the equipment required to recover and recycle the ammonia in the Raschig process. At high production levels the Raschig process is less expensive: the break-even point is 1.8 million pounds of hydrazine per year. For large quantities, hydrazine can be produced by the Raschig process for less than 50 cents per pound (1966 prices). Hydrazine capacity in the USA was about 35 million pounds per year in 1966.

The lack of demand for hydrazine in large quantities has apparently discouraged research and process development to a point where it does not compete as a potential fuel because of its high production costs.

A design study conducted by the Olin Mathieson Company indicated the possibility of producing hydrazine to sell for 20 cents per pound using the Raschig process. However, this required a large installation (10^3 ton/day) with optimisation of operation variables, utilisation of water, heat, etc.

Other processes for making hydrazine have been proposed and demonstrated, but as yet none appear practical. Small quantities of hydrazine are produced in the Haber process for ammonia production. Thus it would appear that a modification of this process might give good yields of hydrazine.

However, since hydrazine undergoes decomposition with energy release, its formation must be accomplished under relatively mild conditions. The successful development of a hydrazine process along these lines should yield a product in the same cost range as ammonia.

Projecting into the future, hydrazine would have to be synthesised from its elements to stand any chance of competing with ammonia as a synthetic fuel. Considerable effort would therefore be required to commercialise such an operation.

Again, extreme toxicity of the fuel means that hydrazine can only be considered as an effective energy carrier or energy storage system. A quick glance at the properties of hydrazine will show that it is marginally a more effective storage medium than ammonia but such an advantage is strongly outweighed by the low potential for any technological breakthroughs in production.

Methanol

Methanol is a high volume chemical produced from 'synthesis gas' composed of CO, H_2, and CO_2. The synthesis gas is generally produced from selected petroleum fractions, natural gas, or in some cases from by-product gas streams. Although methanol has a relatively low heat of combustion, its physical characteristics are such that it would seem to be a portable fuel of wide potential application.

Methanol has been demonstrated to be an attractive fuel for industrial boiler service and then could also be used for conventional electric power generation. Methanol has been used in fuel cells both directly and reformed to hydrogen, and a substantial development programme is currently in progress in France (Esso at Alstholm) to apply methanol to automotive fuel cells. It has also been evaluated as a gas turbine fuel showing several beneficial characteristics including reduced NO_x emission and lower maintenance costs. For example, indications are that the operating time between major maintenance will be some 80 000 hours, eight times that attained using conventional jet fuels. Several studies of the use of methanol as a fuel in conventional internal combustion engines show that it possesses considerable potential for a combination of improved engine performance and reduced emission of unburned fuels, CO, and NO_x. Work on an engine test stand showed methanol to produce, in comparison to gasoline, without catalytic treatment of exhaust gases about 1/20 the unburned fuel, 1/20 the CO, and about the same NO_x. Another experimental study used road performance and test stand measurements of a methanol-fuelled Gremlin engine equipped with a catalytic exhaust reactor to obtain the results given below. Data are also given on the proposed Federal Standards for internal-combustion engine gas emission which were implemented in 1975–1976, and on the emission from Gremlin gasoline-fuelled engine.

$$\text{Emission/g mile}^{-1}$$

	UHC	CO	NO_x
1975–1976 US Federal Standards	0.41	3.4	0.40
Gasoline	2.34	22.08	4.01
Methanol	0.69	3.83	0.28

Methanol shows promise as a long-term, high-quality, domestically produced portable fuel, particularly as a replacement for gasoline.* Since it is basically a high-octane, clean-burning fuel (without additives) it can be used in high-performance engines with apparently relatively simple exhaust treatment, and this may compensate, in part, for its low heating value.

Since the production of methanol is accomplished by reacting the most basic components (CO, CO_2, and H_2) it is not reasonable to assume that future developments would greatly improve the present process except for savings which might be realised by going to huge plant sizes.

There appear to be many sources of CO and CO_2 available as either a by-product or a primary product for many years to come. For the very long-term, however, obtaining a source of carbon as required for the production of methanol would represent a relatively costly process. Possible sources include CO_2 available from the atmosphere (0.03 %), from seawater (as bicarbonate) and from limestone.

Whilst methanol appears as a real competitor to synthetic natural gas (SNG), the fact that some carbon source is required for its synthesis means that the longer term prospects are not so promising.

Ethanol
Industrial-grade ethanol is produced at a rate of about 10^6 tons per year from ethylene derived from crude oil. Production via the fermentation route only accounts for some 10 % of production.

Although ethanol appears to fulfil most portable fuel requirements, high costs have historically limited its application as a fuel. The long-term picture for ethanol appears more hopeful since the fuel can be considered a form of solar energy, produced via photosynthesis and fermentation of grains, sugars, or starches, methanol from waste products, and wood from tree farms. The technology associated with crop processing is certainly well developed but in

---

* After what seems to have been a successful series of studies it appears that Poland may become the first country to adopt the widespread use of methanol + gasoline mixtures in cars. From May 1975 to May 1976 a series of road tests were carried out at the Plock petrochemicals complex using a gasoline + methanol mixture to power the Polski Fiat. Signs now suggest that Poland is examining possible major expansions to its methanol complex at Chorzow to provide fuel-grade product. Poland has the technology to build 2000 tons per day methanol plants, and with very large coal reserves (and claims to have 'the world's best catalyst'—Cu, Zn, Al) for medium pressure synthesis of methanol from carbon dioxide, carbon monoxide and hydrogen.[54]

the past the overall economics have not been favourable due to the low cost of competing fossil fuels. This situation obtains no longer.

Waste processes

It is being increasingly realised that much of the energy tied up in fabricated materials can be made available by judicious processing of waste. In the USA, for example, there occurs about 3×10^9 tons of solid organic waste every year, of which about 80% is of agricultural origin. Municipal and industrial wastes are about 0.4×10^9 tons, and half of this is collected by municipal refuse collectors.[51] The US National Bureau of Mines[52] has pointed out that more than half the total weight of these wastes is water, and in 1971 the total amount of collectable dry organic wastes was 136 million tons. Nonetheless, this amount could yield 170 million barrels of oil (equivalent to 3% of 1971 consumption of crude in the USA or 12% of imported crude). This amount of waste could alternatively have produced 1.36×10^{12} SCF of methane (about 6% of the USA's consumption of natural gas in 1971). If all the wastes were collected purely for methane production then this could yield almost 40% of the USA's natural gas requirements (for 1971).

CHAPTER 7

COAL—THE FOSSIL FUEL OF THE FUTURE

In chapter 2 it was pointed out that one of the oldest methods of producing hydrogen (as a constituent of 'town gas') is from coal. Moreover, table 5.2 shows that coal conversion to liquid fuels requires 6000–7000 SCF of hydrogen per barrel of synthetic oil and about 1500 SCF to produce 1000 SCF of synthetic gas. Thus, it is seen that there can be an intimate connection in production/use of both hydrogen and coal. Presently the industrial world runs on a hydrocarbon economy and there is an immense investment (economic ≡ energy) in this hydrocarbon economy. With the huge reserves of coal (see chapter 1 and figure 7.1) available it is imperative that large-scale conversion to 'clean' hydrocarbon fuels be exploited. In order to do this the industrial production of hydrogen will have to be considerably expanded, and in this way one might see the dawn of the hydrogen era in about 50–75 years when production methods from solar and/or nuclear sources are well developed and our economy has become better adjusted to the change, slight though it may be, from synthetic hydrocarbons to hydrogen.

It is thus part of my thesis that there will be a considerable extension of the fossil fuel era, and that coal is the fossil fuel of the future. It is beyond dispute that in a hundred years from now energy will be transported via a synthetic fuel, and it appears to me that because of the immense amounts of energy involved, and because of a number of environmentally important factors, the transport medium will generally be hydrogen rather than electricity. For such a change to come about preparation is needed. Fortunately research and development in the USA is off to a good start, and the foundation of the Gas Research Institute (to perform similar functions as has the Electric Power Research Institute) in 1976 is symptomatic of the serious intentions of the Americans in planning for large-scale production of synthetic gas from coal, and ultimately of hydrogen.[54] The prognosis for the future is quite bright, thanks to coal, since the return to a coal based economy will allow time and energy for the development of other more permanent energy sources.

The subject of coal as an energy source cannot be dealt with adequately here. However, the point should be made that coal is not an attractive fuel when burned simply as coal; most of the world's coal supplies have unacceptable sulphur contents (see reference 55 for a discussion of acid rains), and so this chapter will deal essentially with coal gasification and liquefaction, emphasising (table 5.2) that exploitation of coal in this way requires sizeable amounts of hydrogen. It also needs to be pointed out that research into ways of winning coal needs to be speeded up, and the political problem of recognising the importance of coal in the long term and persuading workers to join the industry is a factor of importance in the UK if not in the USA. Moreover, transport of coal, especially over long distances as in the USA, needs further development and investment. The Black Mesa coal-slurry pipeline, which is owned by the Southern Pacific Railroad, supplies all the fuel for two 750 MW power stations, transporting 5.5×10^6 tons of coal per year in an 18 inch diameter pipeline (this amount would need 160 rail cars per day). It is thought possible to construct 1000 mile pipelines with capacities of up to 30×10^6 tons of coal per year, but such systems may not be feasible because of the limitation of water resources (especially in the Western USA).[56]

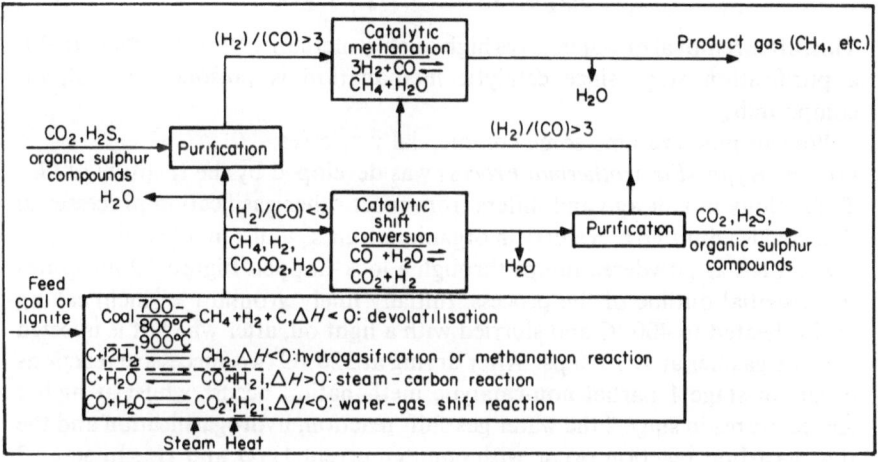

Figure 7.1 General process scheme for producing methane from coal

Coal gasification
The essential features of a gasifier are shown in figure 7.1. Coal is initially heated with steam and, as seen in the figure, a series of reactions take place under non-isothermal conditions. Coal is volatilised

$$\text{Coal} \xrightarrow{600\text{–}900\,°\text{C}} CH_4 + C + H_2$$

and heat is evolved. The carbon in turn reacts with hydrogen in the exothermic hydrogasification (or methanation) step

$$C + 2H_2 \rightleftharpoons CH_4$$

and in step with this is the endothermic steam–carbon reaction

$$C + H_2O \rightleftharpoons CO + H_2$$

Since coal compositions vary the subsequent concentrations of CO, H_2O, CO_2, and H_2 might be thought to vary, but they are almost maintained in equilibrium by the water gas equilibrium

$$CO + H_2O \rightleftharpoons CO_2 + H_2$$

If at this stage the H_2/CO mole ratio is less than 3 the gas mixture is subjected to catalytic shift conversion

$$CO + H_2O \rightleftharpoons CO_2 + H_2$$

in order to increase the hydrogen concentration. When the mole ratio $H_2/CO > 3$ catalytic methanation is performed

$$3H_2 + CO \rightleftharpoons CH_4 + H_2O$$

This, after removal of water, gives high quality methane. Note that there is also a purification stage since catalytic methanation is poisoned by sulphur compounds.

We can now examine some commercial processes.

(a) *The Hygas-Electrothermal Process* was developed by the Institute of Gas Technology in Chicago and differs from most other gasification processes in that coal is admitted as a slurry in organic solvents, whilst in other processes it is admitted as powdered lumps through a lock-hopper. Figure 7.2 illustrates the essential outline of the process. Initially finely ground ($<\frac{1}{8}$inch) caking coal is heated to 400 °C and slurried with a light oil, after which it is injected into the gasifier at $\approx$ 1500 psi. After drying at 320 °C two successive reactions occur. In stage 1 partial non-catalytic methanation occurs whilst at higher temperatures in stage 2 the water gas shift reaction, hydrogasification and the steam–carbon reaction occur with countercurrent H_2O and H_2. In stage 2 some electrical heating energy is supplied, but by keeping the amount of methanation to a maximum this exothermic reaction supplies most of the energy required to drive the endothermic steam–carbon reaction. For gas mixtures where $H_2/CO > 3$ catalytic methanation then follows.

(b) *The Lurgi Process* is that developed by the West German Lurgi-Gesellschaft fur Mineralotechnik GmbH and Badische Anilin-und Soda-Fabrik AG, and has been in use for about 35 years, and is a highly efficient system making use of about 95% of the fuel energy available. In figure 7.3 it may be seen where the small briquette coal ($\sim\frac{1}{8}$inch) enters through the coal lock hopper. After devolatilisation the coal is gasified with steam, heat being supplied from the exothermic carbon combustion. An international conglomerate has recently (1973) built a small Lurgi reactor in Westfield in Scotland in order to discover whether it is possible for methanation to be carried out in a continuous operation using H_2 and CO gases generated from commercial coal.

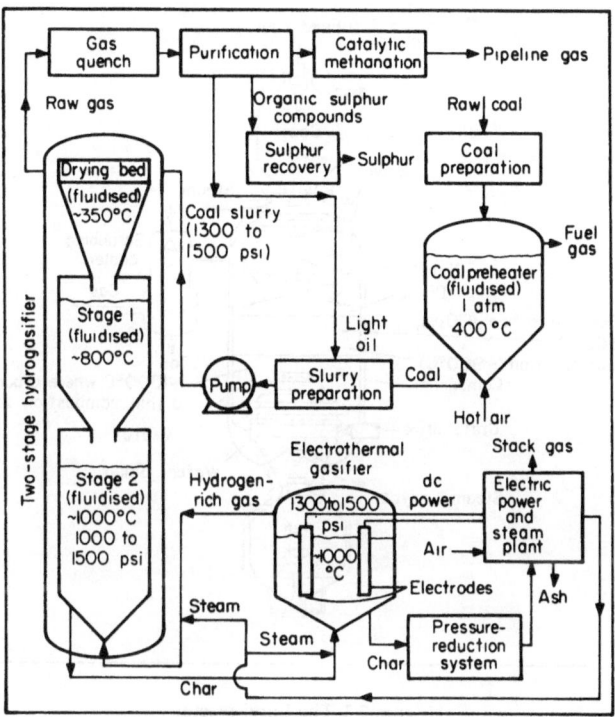

Figure 7.2 Flow diagram for the Hygas electrothermal process

One of the chief advantages of the Lurgi gasifier is its low operating pressure in the gasification chamber (~ 400 psi) (c.f. the Hygas process of > 1000 psi).

The chief problem facing operators willing to commercialise processes in the USA is the ridiculously cheap artificial price of natural gas. Nonetheless, Dr C. A. Stokes, of the Stokes Consulting Group, has recently urged chemical companies to seriously consider the possibility of using gas from coal as a boiler fuel. He predicts that by 1980 gas from coal will be being produced for about $3.50 per million BTU. Stokes knows that it is frustrating to try to commercialise synthetic gas production with current energy economics as they are today in the USA, and he knows that gas from coal will not be economically attractive until the US government decontrols natural gas prices. He says[58]

As a nation, we struggle to use coal to make fuel gas when natural gas . . . sells to contract industrial consumers at $1.25–1.75 per million BTU. Only when natural gas is allowed to compete on the basis of the cost to replace it ($3.50–4.00 per million BTU) will coal-based gas really begin to move.

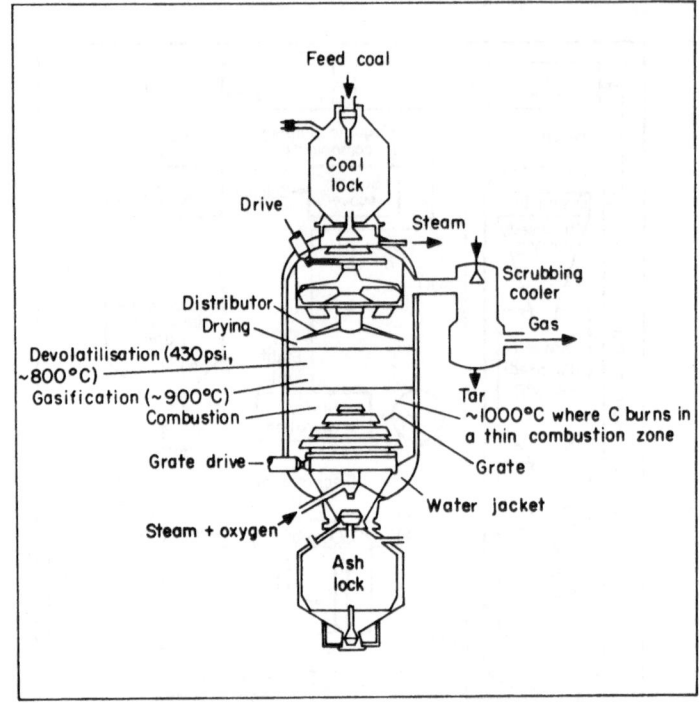

Figure 7.3 The Lurgi reactor

One must hope that short-term commercial interests are overridden for the long-term good and that the US government will soon sensibly price natural gas.

Table 6.2 contains some production costs of energy contained in selected fuels.

Coal liquefaction
It is this aspect of the use of coal where the production of hydrogen becomes a necessity.

In 1955 the South African Coal, Oil, and Gas Corporation began a process (the SASOL process) which was essentially similar to procedures used in wartime Britain to produce synthetic crude oil. Coal is initially gasified in a Lurgi reactor with steam, oxygen, and CO_2, and sulphur compounds are removed in the usual way. The gas is then heated in the presence of a powdered iron catalyst and on condensation and subsequent distillation liquid hydrocarbons are obtained. Present production in South Africa is about 4000 bl/d of gasoline and 100 bl/d of diesel fuel.

Table 7.1 gives some idea of the cost of oil from coal in 1972. At that time the

Table 7.1 Oil from coal

	Oil yield (lb/ton coal)	H_2 needed (standard ft^3/lb oil)	Cost ($/ton oil)
Liquid phase hydrogen of high volatile coal 10 000 lb/in^2	1220	35	?
Liquid phase hydrogen of high volatile coal 3000 lb/in^2	1080	30	85
Lurgi process, hydrogen of tar into light oil	460	16	50
Fischer–Tropsch gasification	700	0	90

oil produced by the Lurgi process cost about $7 per barrel ($50 per ton) and was expensive. At the beginning of 1977 when oil cost $12–14 per barrel it made economic sense to begin large scale exploitation of coal reserves in this way. (SASOL gasoline, in 1979, cost approximately $120/ton.[59a])

In theory 'syncrude' from coal needs about 5×10^3 SCF of H_2/bl for the removal of unwanted sulphur, oxygen, and nitrogen compounds in order to produce petroleum containing about 13 % by weight of hydrogen, but in practice about 6×10^3 to 10×10^3 SCF are needed.[59]

There are other pilot plant processes curently being examined,[60–62] and one should also emphasise here that it is vital to the world's chemical industry, just as it is to the transport and space-heating industries, that a supply of hydrocarbons be maintaind after the world supply of hydrocarbon fossil fuels has run out. The Ralph M. Parsons Co. of Pasadena, California, has plans for a coal/chemicals complex which would use 66 000 tons of coal per day. The 17 major products to be derived will include one billion pounds per year of ethene, 434 million pounds of propene, 34 million gallons per year of benzene, 16 million gallons of toluene, 71.5 million gallons of mixed xylenes, 2400 tons per day of sulphur, and 214 tons per day of ammonia. Parsons calculates that a little more than three of these complexes could supply 10 % of USA's ethene needs in 1980.[63]

The future
The exploitation of coal reserves would certainly be enormously aided by underground *in situ* gasification (see reference 64 for a recent appraisal of the methods used), a method of using coal reserves first suggested in 1868 by William Siemens. In the Soviet Union investigation of this problem was active between 1933 and 1965, and work has also been prosecuted in the UK, France, Belgium, Italy, and is now once again active in the USA.[65]

In concluding this book I would like to emphasise my final points by reference to table 7.2, from which it can be seen that the capital investment (for capital investment also read consumption of available energy) of basic energy sources points strongly to the exploitation of coal as making economic common sense.

Table 7.2 Capital costs of basic energy sources (1974)

Oil equivalent 100 000 bl/day production

	£m Output basis	£m Input basis
Nuclear power	1500	500 (75% load factor)
Coal power station	1200	400 (75% load factor)
Tar sands	350	
Oil shale	250–300	
Coal gasification	250	
North Sea oil	150	
Gas recovery and liquefaction	120	
Coal mining	15	

REFERENCES

1. S. S. Penner and L. Icerman, *Energy*, Volume 1, Addison-Wesley (1974).
2. 'Energy and Power', *Scientific American*, a collection of reprints (1973).
3. *Resources and Man*, American Association for Advancement of Science, Washington DC.
4. D. J. Rose, quoted in 'Chemistry and the Environment', Freeman (1973).
5. C. A. McAuliffe, 'The Hydrogen Economy', *Chemistry and Industry*, 725 (1975).
6. *A Hydrogen Energy System*, American Gas Association (1972).
7. J. E. Funk and R. M. Reinstrom, 'Energy Requirements in the Production of Hydrogen from Water', *J.E.C. Proc. Res. and Dev.*, **5**, 336 (1966).
8. G. De Beni, 'Hydrogen Production Cyclic Process', French Patent 2035558, Feb. 11th 1970.
9. G. De Beni, and C. Marchetti, 'Mark I, A Chemical Process to Decompose Water Using Nuclear Heat', paper presented at the 163rd National Meeting of the American Chemical Society, Boston, Mass. (1972).
10. K. F. Knoche and J. Schubert, paper presented in Symposium on Chemical Fuels at 166th National Meeting of the American Chemical Society, Chicago, Illinois (1973).
11. C. Hardy, Euratom Report 49581 (1973).
12. 'Hydrogen Production From Water Using Nuclear Heat', Progress Report No. 3, Euratom 5059e (1973).
13. R. H. Wentorf, and R. E. Hunneman, 'Thermochemical Hydrogen Generation', General Electric Corporation, Schnectady, New York, Report No. 73CRD222 (1973).
14. R. M. Reinstrom, 'Ammonia Production Feasibility Study', EDR 4200 (1965).
15. C. Marchetti, 'Hydrogen and Energy', *Chem. Econ. Eng. Rev.*, **5**, 5 (1973).
16. N. C. Hallert, 'Study Cost and Systems Analysis of Liquid Hydrogen Production', NASA CR 73-226 (1968).
17. W. F. Eagle and M. J. Waale, 'Recent Developments in the Oxidation Recovery of Chlorine from Hydrochloric Acid', *Chem. Ind.*, 76 (1962).
18. A. R. Miller and H. Jaffe, 'Process for Producing Hydrogen from Water Using an Alkali Metal', US Pat. 3490871, 20 January 1970.
19. Progress Report No. 3, Euratom Joint Nuclear Research Centre, Ispra, Italy. Report EUR/c–15/35/75e.

20. W. H. McCulloch, R. B. Pope and D. O. Lee, 'Economic Comparison of Two Solar/Hydrogen Concepts', Sandia Laboratories Report No. SLA–73–0900.
21. G. De Beni, 'Considerations on Iron–Chlorine–Oxygen Reactions in Relation to Thermochemical Water Splitting', THEME Conference Reports, Miami, p. 511–513 (1974).
22. J. B. Pangborn and J. C. Sharer, 'Analysis of Thermochemical Water Splitting Cycles', THEME Conference Reports, Miami, p. 54–55 (1974).
23. 'A Hydrogen Energy Carrier: Volume II. Systems Analysis', NASA–ASEE Systems Design Institute, University of Houston, Johnson Space Center, Rice University Report, p. 39 (1973).
24. G. E. Daniels (ed.), 'Terrestrial Environment (Climatic) Criteria Guidelines for use in Space Vehicle Development 1971 Revision', NASA Technical Memorandum TMX–64589, p. 2.6 (1971).
25. H. Gaffron and J. Rubin, *J. Gen. Physiol.*, **26**, 219 (1942).
26a. H. Gest and M.D. Kamen, *Science*, **104**, 558 (1949).
26b. R. Benneman and D. J. Weare, *Science*, **184**, 174 (1974).
27. G. Istings, 'Pipelines Now Play Important Role in Petrochemical Transport', *World Pet.*, April 1970, p. 40.
28. G. A. Nelson, 'Use Curves to Predict Steel Life', *Hydrocarbon Process*, **44**, 185 (1965).
29. R. P. Jewett *et al*, 'Hydrogen Environment Embrittlement of Metal', A NASA Technology Survey, Rocketdyne Division of North American Rockwell, Final Report on Contract NA 58–10(c), Washington, DC.
30. R. A. Reylonds and W. L. Slayer, 'Pipeline Transmission of Hydrogen', paper presented at THEME Conference, Miami (1974).
31. A. H. Shapiro, *The Dynamics and Thermodynamics of Compressible Fluid Flow, New York*, Ronald Press (1953).
32. J. P. O'Donnell, 'Pipeline Economics', *Oil and Gas.*, **71**, 76 (1973).
33. R. L. Whitelow, 'Electric Power and Fuel Transmission by Liquid Hydrogen Superconductive Pipeline', paper presented at THEME Conference, Miami (1974).
34. D. P. Gregory, 'The Hydrogen Economy', *Scientific American*, **228**, 1 (1973).
35. H. Jones, 'The Storage of Gas in Underground Aquifers', *Int. Gas Eng. J.*, **3**, 257 (1963).
36. M. D. Tade, 'Helium Storage in Cliffside Fields', paper presented at The Society of Petroleum Engineering Regional Meeting, Amarillo Texas (1966).
37. C. J. Kippenhan and R. C. Corlett, 'Hydrogen Energy Storage for Electrical Utility Systems', paper presented at THEME Conference, Miami (1974).
38. American Gas Association, *LNG Information Book*, (1968).

39. N. C. Hallett, 'Cost and Systems Analysis of Liquid Hydrogen Production', Air Products Inc. NASA Report CR73, 226, Washington DC (1968).
40. J. W. Michel, 'Hydrogen and Synthetic Fuels for the Future', paper presented at the American Chemical Society Symposium, Chicago, Illionois (1973).
41. K. M. Mackay, *Hydrogen Compounds of the Metallic Elements*, Spon (1966).
42. K. C. Hoffman, W. E. Winsche, R. H. Wiswall, J. J. Reilly, T. V. Sheehan, and C. H. Waide, 'Metal Hydrides as a Source of Fuel for Vehicular Propulsion', Proceedings of the International Auto Engineering Congress, Chicago (1969).
43. J. J. Reilly and R. H. Wiswall, 'The Reaction of Hydrogen with Alloys of Magnesium and Nickel and the Formation of Mg_2NiH_4', *Inorg. Chem.*, 7, 2254 (1968).
44. J. J. Reilly and R. H. Wiswall, *Inorg. Chem.*, 13, 218 (1974).
45. J. H. N. van Vucht, F. A. Kuijpers, and H. C. A. M. Brunning, *Phillips Research Reports*, 25, 133 (1970).
46. G. D. Brewer, paper presented at THEME Conference, Miami (1975)
47. G. D. Brewer, 'Advanced Supersonic Technology Concept Study – Hydrogen Fueled Configuration', NASA Report CR114, 718, prepared by the Lockheed California Company, (1972).
48. G. D. Brewer, R. E. Morris, R. H. Lange, and J. W. Moore, 'Study of the Application of Hydrogen Fuel to Long-Range Subsonic Transport Aircraft', NASA CR132559, prepared by the Lockheed California Company and The Lockheed Georgia Company (1975).
49. Study of Structural Design Concepts for an Arrow-Wing Supersonic Transport Configuration', NASA–Langley Research Centre report to Lockheed California Company (1973).
50. US National Bureau of Mines Information quoted in reference 6.
51. H. Feldmann, 'Pipeline Gas from Solid Wastes', paper presented at American Institute of Chemical Engineers, Annual Meeting, Cincinnati, Ohio (1971).
52. G. E. Johnson, 'The Production of Ammonia by Anaerobic Decomposition of Garbage and Waste Materials', paper presented at the 163rd National Meeting of the American Chemical Society, Boston, Mass. (1972).
53. *European Chemical News*, 31 December 1976, p. 18.
54. *Chemical and Engineering News*, 16 August 1976, p. 5.
55. *Chemical and Engineering News*, 22 November 1976, p. 29.
56. 'National Academy of Sciences: Water Scarcity May Limit use of Western Coal', *Science*, 181, 525 (1953).
57. J. P. Henry and P. M. Louks, 'An Economic Comparism of Processes for

Producing Pipeline Gas (Methane) from Coal', presented at the September 1970 Meeting of The American Chemical Society, Chicago, Illonois.

58. *Chemical and Engineering News*, 6 September 1976, p. 13.

58a. H. Schulz, *Pure Appl. Chem.*, **51,** 2225 (1979).

59. National Petroleum Council, 'US Energy Outlook: Coal Availability', Washington DC (1973).

60. 'Background Information on Solvent Extraction of Coal', Hydrocarbon Research Inc., Palo Alto, California, Report 123–1–0 (1974).

61. 'Commercial Process Evaluation of the H-Coal Hydrogenation Process', Hydrocarbon Research Inc., US Government Printing Office, Washington, D. C. Report PB–174696 (1967).

62. *Chemical and Engineering News*, 23 September 1974, p. 7.

63. *Chemical and Engineering News*, 7 September 1976, p. 7.

64. R. M. Nadkarni, C. Bliss and W. I. Watson, 'Underground Gasification of Coal', *Chem. Tech.*, April 1974, p. 230.

65. *Chemical and Engineering News*, 3 December 1979, p. 19.

APPENDIX: UNITS AND CONVERSION FACTORS

Although there is a movement in scientific circles to express quantities in SI units, it is still the overwhelming practice of those involved in energy conversion and distribution to use the MKS system, and some conversions are given here in the latter system.

It may also be helpful to give some definitions:

Energy may be expressed as the ability to do work, where *work* is defined as a force acting through a distance parallel to the force, i.e. $W = Fd$. There are basically three types of energy, *kinetic, potential* and *rest-mass. Kinetic energy* is the energy of motion, K.E. $= \frac{1}{2}mv^2$. *Potential energy* occurs in a number of different forms, e.g. solar energy is converted into the potential energy of evaporated water from the earth's surface deposited on higher ground. In this case, P.E. $= mgh$ ($m =$ mass of water in kg, g is the acceleration due to gravity, and h is the vertical distance through which the water is moved). Thus 1 kg of water raised 10 000 m would have a K.E. $= 1 \times 9.8$ ms$^{-2} \times 10\,000$ m $= 98\,000$ km^{2}s$^{-2} \equiv 98$ kJ. *Rest-mass energy* is conceptually the energy which fuels the sun, i.e. a fraction of the mass of hydrogen is converted into heat and radiant energy: $E = mc^2$, where m is the mass converted (in kilograms) and c is the speed of light (3×10^8 m s^{-1}), e.g. for the conversion of 1 kg of reactor fuel in a fusion process the energy released is $1 \times (3 \times 10^8)^2$ J $= 9 \times 10^{16}$ J.

In this book, as in any volume devoted to energy, the concept of *power* must be defined. It is the rate of doing work, i.e. $P = W/T$, where W is usually expressed in joules and T is time in seconds. Thus P has units of joules per second (1 J s$^{-1} = 1$ watt). Thus a 150 watt (150 W) light bulb consumes 100 J/s. Power is also expressed as horsepower (hp), and therefore 1 hp $= 746$ W, and a car of 200 hp is equivalent to 149 200 W (149 kW).

Some important units and conversion factors are:

1 J $= 10^7$ erg (or 10^7 dyne cm)
$ = 6.24 \times 10^{12}$ MeV
$ = 0.1$ N m $= 0.736$ ft lb
$ = 0.24$ cal $= 0.949 \times 10^{-3}$ BTU
$ = 2.78 \times 10^{-4}$ W h $= 3.73 \times 10^{-7}$ hp h

$= 1.22 \times 10^{-13}$ of fusion energy from the deuterium contained in 1 m^3 of seawater

$= 1.11 \times 10^{-4}$ g of matter equivalent

$= 1.22 \times 10^{-14}$ of fission energy of 1 kg of U-235 equivalent

$= 6.7 \times 10^{-23}$ of incident solar energy on the earth's atmosphere

$= 5.8 \times 10^{-32}$ of daily solar output

$1 \text{kW h} = 8.62 \times 10^5 \text{ cal} = 3.6 \times 10^6 \text{ J} = 3.41 \times 10^3 \text{ BTU}$

$\qquad = 3.41 \times 10^{-2} \text{ therm} = 2.65 \times 10^6 \text{ ft lb}$

$\qquad = 2.25 \times 10^{25} \text{ eV}$

$\qquad = 6.06 \times 10^{-1} \text{ bl oil}$

$\qquad = 2.6 \text{ lb bituminous coal}$

$\qquad = 3.2 \text{ SCF natural gas}$

1 metric ton of coal $\simeq 27.8 \times 10^6$ BTU

1 bl oil $\simeq 5.7 \times 10^6$ BTU

1 SCF natural gas $\simeq 10^3$ BTU

$1 \text{ Q} = 10^{18} \text{ BTU}$

$\qquad = 1.05 \times 10^{21} \text{ J}$

$\qquad = 2.93 \times 10^{14} \text{ kW h}$

$\qquad \simeq 1.7 \times 10^{11}$ bl oil equivalent

Coal conversion to oil: $\geqslant 2$ bl/mt

INDEX